Darwinism and Natural Theology

Darwinism and Natural Theology: Evolving Perspectives

Edited by

Andrew Robinson

CAMBRIDGE
SCHOLARS

PUBLISHING

Darwinism and Natural Theology: Evolving Perspectives,
Edited by Andrew Robinson

This book first published 2012

Cambridge Scholars Publishing

12 Back Chapman Street, Newcastle upon Tyne, NE6 2XX, UK

British Library Cataloguing in Publication Data
A catalogue record for this book is available from the British Library

ISBN (10): 1-4438-3742-3, ISBN (13): 978-1-4438-3742-2

CONTENTS

ACKNOWLEDGEMENTS

The Science and Religion Forum is grateful to the Principals and staff of Wesley House and Westcott House, Cambridge, for hosting the conference of which these chapters are the product. Their hospitality and the congenial environment in which the conference took place contributed greatly to the quality of the scholarly interactions that lie behind this volume.

I wish to record my own thanks to Neil Spurway for sharing his editorial expertise and making some vital contributions to the editorial process, and to Charles Robinson for assisting with the references and bibliography. I am also most grateful to Carol Koulikourdi, Amanda Millar and Soucin Yip-Sou of Cambridge Scholars Publishing for their work in bringing the volume to publication.

Andrew Robinson, April 2012.

CONTRIBUTORS

Denis Alexander is Director of The Faraday Institute for Science and Religion at St. Edmunds College, Cambridge, where he is a Fellow. He was previously at The Babraham Institute where he was Chair of the Molecular Immunology Programme and Head of the Laboratory of Lymphocyte Signalling and Development. He is Editor of the journal *Science and Christian Belief*, serves on the national committee of Christians in Science and is a founding member of the International Society for Science and Religion. His recent books include *The Language of Genetics – an Introduction* (Templeton Foundation Press, 2011), co-editor with Ronald Numbers of *Biology and Ideology - From Descartes to Dawkins* (Chicago University Press, 2010), *Creation or Evolution - Do We Have to Choose?* (Monarch, 2008), editor of *Can We Know Anything? Science, Faith and Postmodernity* (Apollos, 2005), and co-author (with Robert White FRS) of *Beyond Belief - Science, Faith and Ethical Challenges* (Lion, 2004).

R.J. (Sam) Berry was Professor of Genetics at University College London from 1984-2000. He is a former President of the Linnean Society, the British Ecological Society, the European Ecological Federation, the Mammal Society, and Christians in Science. He is a Vice-President of the Science & Religion Forum, the John Ray Initiative and the Field Studies Council. In 1996 he received the UK Individual Templeton Award for his "sustained advocacy of the Christian faith in the world of science." He gave the Gifford Lectures in the University of Glasgow 1997-98 (published as *God's Book of Works*, T&T Clark, 2003). His most recent books are *Islands* (Collins, 2009) and *Ecology and the Environment: the Mechanisms, Marring and Maintenance of Nature* (Templeton Press, 2011); in collaboration with Michael Northcott he edited *Evolution After Darwin* (Paternoster, 2009) and with Tom Noble *Darwin, Creation and the Fall* (Apollos, 2009).

John Hedley Brooke held the Andreas Idreos Chair of Science & Religion and Directorship of the Ian Ramsey Centre at Oxford University from 1999 to 2006. He is an Emeritus Fellow of Harris Manchester College Oxford, Honorary Professor of the History of Science at Lancaster

University and, in 2007, was appointed "Distinguished Fellow" at the Institute of Advanced Study, University of Durham. From 2008 to 2011 he was President of the International Society for Science and Religion and is currently President of the UK Science and Religion Forum. His books include *Science and Religion: Some Historical Perspectives* (Cambridge University Press, 1991) and (with Geoffrey Cantor) *Reconstructing Nature: The Engagement of Science & Religion* (T&T Clark, 1998). His latest book, co-edited with Ronald Numbers, is *Science & Religion Around the World* (Oxford University Press, 2011).

Michael Burdett is a Junior Research Associate at the Oxford Centre for Christianity and Culture, Regents Park College, University of Oxford, and Associate Tutor at Wycliffe Hall. He holds degrees in engineering, physics, and theology and has been given academic and professional awards in each field. Currently, he is working on a project funded by the John Templeton Foundation which researches the resourcing of science and religion at Christian universities in the USA. He has published and presented internationally on continental philosophy, the technological society and Christian theology.

Philip Chapman's chemistry degree at Bristol in 1954-57 included diversions to the philosophy and history of science with lectures from Paul Feyerabend on Karl Popper. Subsequently he studied theology at Birmingham, and, much later, at Leeds Centre for Science and Religion with Dr Jacqui Stewart. While studying French in Paris in 1964 he met Karl Barth. Forty years active Methodist ministry included fourteen years, far from academia, in the tropical forest of francophone West Africa. After retirement, while for a year in pastoral charge of churches among Sri Lankan tea estates, a member suggested that he should work for a PhD in theology, which he is now doing under Professor Neil Messer at Winchester University.

Sarah Coakley is the Norris-Hulse Professor of Divinity at the University of Cambridge, and was previously Mallinckrodt Professor of Divinity at Harvard University, where she undertook a three-year joint research programme with the evolutionary biologist and mathematician Martin A. Nowak. Amongst her publications are *Christ Without Absolutes* (Oxford University Press, 1988), *Powers and Submissions* (Blackwell, 2002), *Pain and Its Transformations, co-ed.* (Harvard University Press, 2008), and *God, Sexuality and the Self* (Cambridge University Press, forthcoming,

2012). She will deliver the 2012 Gifford Lectures at Aberdeen on *Sacrifice Regained: Cooperation, Evolution and God.*

Celia Deane-Drummond is currently full Professor in Theology at the University of Notre Dame, USA, concurrent between the College of Arts and Letters and the College of Science. She holds doctorates in plant physiology and in theology. Her books include *Creation through Wisdom* (T&T Clark, 2000), *Brave New World* (Continuum, 2003), *Reordering Nature* (Continuum 2003), *The Ethics of Nature* (Blackwell, 2004), *Wonder and Wisdom: Conversations in Science, Spirituality and Theology* (DLT, 2006), *Genetics and Christian Ethics* (Cambridge University Press, 2006), *Future Perfect: God, Medicine and Human Identity*, ed. with Peter Scott (Continuum, 2006, 2nd edn. 2010), *Ecotheology* (DLT/Novalis/St Mary's Press, 2008), *Christ and Evolution: Wonder and Wisdom* (Fortress / SCM Press, 2009), *Creaturely Theology: On God, Humans and Other Animals*, ed. with David Clough (SCM Press, 2009), *Seeds of Hope* (CAFOD, 2010), *Religion and Ecology in the Public Sphere*, ed. with Heinrich Bedford-Strohm (Continuum, 2011), and *Rising to Life*, ed. (CAFOD, 2011).

David Fergusson is Professor of Divinity and Principal of New College at the University of Edinburgh. He is the author of *Faith and its Critics* (Oxford University Press, 2008), based on the 2008 Gifford Lectures delivered in the University of Glasgow. He is also editor of the *Blackwell Companion to Nineteenth Century Theology* (Wiley-Blackwell, 2010) and co-editor of the *Cambridge Dictionary of Christian Theology* (Cambridge University Press, 2011). He is a Fellow of the Royal Society of Edinburgh.

David Knight has taught History of Science at Durham University since 1964, where he was chair of the Philosophy Department in the 1990s. His module on Science and Religion in the nineteenth century won awards from the Templeton Foundation. He is a past editor of *The British Journal for the History of Science*, and past President of the British Society for the History of Science. He is now doing his fourth stint as a churchwarden of St Oswald's Church, Durham. He has played many parts, including God, Reason, Death, Prospero, Justice Shallow and Snug the joiner. He wrote a number of entries for The Oxford Dictionary of National Biography, 2004, and among his books are *Science and Spirituality: the Volatile Connection* (Routledge, 2004), and *The Making of Modern Science: Science, Technology, Medicine and Modernity,1789-1914* (Polity Press, 2009).

Fabien Revol is an associate professor at the Interdisciplinary Centre for Ethics at the Catholic University of Lyon. He holds degrees in the biology of ecosystems and in theology. He was awarded the European Society for the Study of Science and Theology (ESSSAT) student prize in 2008. He is currently working on his doctoral thesis on science and religion with special reference to the link between the concept of continuous creation and the evolution of life, and is developing a new curriculum in bioethics for the Faculty of Science of his university. He is co- editor of *Evolution et Création: des sciences à la métaphysique* (Vrin IIEE, 2011).

Andrew Robinson is a general medical practitioner in Newton Abbot, Devon, and Honorary University Fellow in Theology at the University of Exeter. He holds degrees in physiology and medicine from Bristol University and a doctorate in theology from Exeter. He is author of *God and the World of Signs: Trinity, Evolution and the Metaphysical Semiotics of C.S. Peirce* (Brill, 2010). He is a member of the Royal College of Physicians and the Royal College of General Practitioners, and in 2011 was elected to membership of the International Society for Science and Religion.

Christopher Southgate trained initially as a biochemist at Cambridge. He is a poet, editor and theologian, and is currently Research Fellow in Theology at the University of Exeter, and Vice-Principal of the South-West Ministry Training Course. His most recent monographs are *The Groaning of Creation: God, Evolution and the Problem of Evil* (Westminster John Knox, 2008) and the co-written *Greening Paul: Re-reading the Apostle in a Time of Ecological Crisis* (Baylor University Press, 2010). He is also the editor of the science-and-religion textbook *God, Humanity and the Cosmos* (T&T Clark, 3rd edn. 2011). Chris is currently exploring the theme of glory, and also pursuing collaborative research on the origin of life – both themes that stem out of the work outlined in the chapter with Andrew Robinson in this volume. Southgate's sixth poetry collection, *A Gash in the Darkness* (Shoestring) is published in April 2012.

Neil Spurway avidly read the Gifford lectures as VIth former. He studied physiology and physics at Cambridge, 1957-60 and continued to a PhD in membrane biophysics (but incurred disapproval for playing truant to philosophy and theology lectures, and editing an SCM journal). Attracted by mountains and sea to Glasgow in 1963, he served over 30 years on the Gifford Lectureships committee. He retired as Professor of Exercise

Physiology in 2001 and has since been Chair of the British Association of Sport and Exercise Sciences, President of the Royal Philosophical Society of Glasgow, journal Editor and Vice-President of European Society for the Study of Science and Theology, and Chair of the UK Science and Religion Forum. At 75, he sails his boat for several months each summer to different countries of northern Europe, but in autumn and winter still teaches courses in physiology, philosophy of science and science-and-religion. His particular interest is in evolutionary epistemology and its implications for theology.

Kenneth Wilson held the Rowbotham Chair in philosophical theology and ethics at Wesley College, Bristol from 1973-1981 and lectured in the Faculty of Theology of the University of Bristol. From 1981-1996 he was Principal of Westminster College, Oxford. He established the Research Centre for Theology and the Social Sciences at the Queen's Foundation for Ecumenical Theology, Birmingham, and was the Director from 1996-2001. He was a member of a research programme on 'Authority and Governance in the Roman Catholic Church' which led to the Durham Conference on 'Receptive Ecumenism' in 2006 and an international research programme, 'Ecclesiological Investigations'. He was a founder member of the Ian Ramsey Centre for Science and Religion in the University of Oxford, and is currently the Chair of the UK Science and Religion Forum. He is involved in 'Learning for Life' a research programme focused upon character education. Recent publications include *Learning to Hope* (Epworth Press, 2006), *Dying to Live* (Epworth Press, 2008) and *Methodist Theology* (T&T Clark, 2011).

CHAPTER ONE

INTRODUCTION

ANDREW ROBINSON

The year 2009 saw the 200th anniversary of the birth of Charles Darwin (12 February 1809) and the 150th anniversary of the publication of his *Origin of Species* (November 1859).[1] Various events and conferences marked this double cause for commemoration and reflection, one of the largest being the Darwin Festival hosted by the University of Cambridge in July of the bicentennial year. In one session, four eminent scholars of the field of science and religion reflected on the ways in which theology is affected by, and may thrive within, the Darwinian context.[2] The large lecture theatre was packed to overflowing – evidence itself of the level of interest in the relationship between Darwinism and religion. During questions from the audience a memorable incident occurred. From almost the back row of the steeply banked seating an imposing figure with a Darwinesque beard announced himself as "one of the four horsemen of the apocalypse." The horseman was the philosopher and 'new atheist' Daniel Dennett. He went on to say that the new atheists are often criticized for not engaging sufficiently with the insights and subtleties of real academic theology. But, he continued, he could not tell from anything he had heard that afternoon that theology had anything concrete to offer him by way of additional understanding the world – anything comparable to the insights

---

[1] Charles Darwin, *On the Origin of Species by Means of Natural Selection, or the Preservation of Favoured Races in the Struggle for Life* (London: John Murray, 1859).

[2] "Theology in a Darwinian Context," *Darwin Festival* conference session, Cambridge, 6 July 2009. The session was chaired by Professor Sarah Coakley. The speakers were Philip Clayton, Denis Alexander, Fraser Watts, and J. Wentzel van Huyssteen.

available from science – and he therefore remained unrepentant about his alleged underestimation of the relevance of the discipline.[3]

Dennett's apocalyptic intervention was very much in my mind when, two months later, the Science and Religion Forum convened in Cambridge for its own annual conference. The Forum had, naturally, chosen to link the conference with the Darwin anniversaries. However, conscious, of the fact that many of the big issues arising from Darwinism would already have been pored over in other settings during the year, the Forum had decided to focus its conference on a specific theme. The area it had chosen to discuss was the place of 'natural theology' in the light of Darwinian evolutionary biology. Although planned many months previously, it could not have chosen a better context in which to test Dennett's assertion that theology has no traction with real knowledge of the world. For it must be acknowledged that part of the reason that his intervention at the Darwin Festival was memorable was that his remarks stung: there is indeed a legitimate question about whether the way in which theology engages with Darwinism amounts to anything more than a set of purely defensive and rather desperate moves.

To understand how this book may contribute to responding to that challenge we must consider briefly the meaning(s) of the term 'natural theology'. Natural theology is normally contrasted with 'revealed theology'. Revealed theology is knowledge of God derived from God's special self-revelation in Scripture, in special acts of providence, and for Christians preeminently in the life, death and resurrection of Christ.[4] Natural theology, on the other hand, is knowledge of God derived from reflection on the ordinary working of the natural world. The question arises, how are these paths to knowledge of God related? Classically they are thought to operate somewhat independently of each other. According to this view, natural theology is exemplified by the standard arguments for

---

[3] This incident is also recalled by Sarah Coakley in Chapter 7 (p. 97). If my recollection of it is imperfect I can only apologise and note that such inaccuracy would stand in a venerable tradition of dubious reporting of Darwinian debates (I think, of course, of the doubts surrounding the canonical accounts of the Huxley-Wilberforce exchange at the 1860 meeting of the British Association for the Advancement of Science).

[4] Although the conference at which the papers in this volume were presented focussed on Christian theology, the Science and Religion Forum encourages dialogue between enquirers of all faiths or none. This engagement with a variety of faith traditions is reflected in previous volumes arising from the Forum's conferences. See, for example, Neil Spurway, ed., *Creation and the Abrahamic Faiths* (Newcastle: Cambridge Scholars Publishing, 2008).

the existence of God (paradigmatically in Aquinas's 'five ways', though the extent to which Aquinas himself regarded these as operating independently of revealed theology is a matter of debate). Thus arguments such as the argument from a series of causes to an uncaused cause, or the argument from the regularities of natural processes to a designer outside the world, yield conclusions about the probability of the existence of God. However, whilst such arguments point to the existence of God they cannot, it is often said, provide knowledge of the nature or essence of that God; such knowledge is available through revelation but not through reason.

This understanding of natural theology – as an enterprise that operates independently of revealed theology – has been extensively criticized. Philosophical criticisms of natural theology can be traced to Hume and Kant and rest on criticisms of the premises from which deductive or inductive arguments for the existence of God begin, or on criticisms of the legitimacy of the process of such deductive or inductive reasoning. For example, Hume argued against the reliability of inductive reasoning: from the fact that many *known* As (e.g., instances of order) are caused by Bs (designers) we cannot conclude that *every* A is caused by a B. And the unreliability of inductive reasoning is especially to be guarded against when we consider that invoking a human designer as the cause of a humanly designed artifact is something of which we can give many examples, whereas invoking God as the designer of the world as a whole is (if it is true) a unique relationship of which we have no other comparable examples.

Criticisms of natural theology have also come from within theology itself, notably in the twentieth century from the Reformed theologian Karl Barth. The burden of theological criticisms of classical natural theologies is that they imply that humans have some capacity to know God that is independent of God's gracious self-revelation, and hence deny the absolute sovereignty and transcendence of God. Furthermore (and this is part of Barth's concern, writing as he did in the context of a church all too accepting of Nazism) by seeking knowledge that potentially circumvents the message of Scripture and the incarnation of God's Word in the person of Jesus our theologies may easily be contaminated by something alien to the gospel.

Philosophical and theological criticisms of classical natural theology must clearly be taken seriously. The extent to which such criticisms should be considered decisive is beyond the scope of the present remarks, and indeed beyond the scope of this volume. Rather, the issue that lurks behind the chapters that follow is, in effect, whether natural and revealed theology are in fact properly regarded as distinct and independent paths to knowledge of God. To put the matter rather simply, is it not the case that revealed theology always depends on the operation of some elements of reason and experience that lie outside, say, scripture? Would it be possible to read and understand scripture entirely in isolation, without any prior experience or knowledge of the world? If not, it must surely be the case that some kind of 'natural' knowledge is required before we can receive the revelation contained in scripture. And why should the capacity for such 'natural' knowledge not be regarded as being as much a gift of the absolutely transcendent Creator as that of scripture itself? Conversely, is it possible to have any 'natural' knowledge of the world without some prior metaphysical (including, but not necessarily limited to, religious) commitments that shape such knowledge? Is the mechanistic, atheistic world-view of the new atheists any more free of metaphysical commitments than a theology of nature which takes the Christian gospel as its overarching framework? In short, aren't natural theology and revealed theology much more mutually dependent than the standard view would tend to imply? If so, Christian theology may have more to learn from reflection on the 'natural' world than the theological critics of natural theology tend to grant. But similarly, agnostic or atheist explorers of the 'natural' world may have more to gain from examining their metaphysical presuppositions than they suppose, and possibly something to learn from a specifically Christian reading of the book of nature.

The present volume brings together the papers presented at the Science and Religion Forum's conference in Cambridge in the Darwin bicentennial year. It offers historical, philosophical and theological perspectives on the ways in which Christian theology has been, and may continue to be, generatively informed by engagement with the understanding of the world arising from Darwinian evolutionary biology. It includes a call for theologians to be bolder in their expectations of the possibilities for natural theology positively to shape the metaphysical presuppositions of working scientists, and some examples of such an emboldened natural theology in operation.

Such a nuanced and potentially enriched view of the relation between 'natural' and 'revealed' theologies may, admittedly, not be sufficient to unseat the four horsemen of the apocalypse. But perhaps, in light of this evidence of the vitality of the interaction between Christianity and Darwinism, those knights of the 'new atheism' would be well advised to slow their thundering gallop to a more circumspect trot.

* * *

The volume is divided into four parts. Part 1 is mainly concerned with historical background – with Darwinism and with the variety of positions taken up in response to Darwin's work by those who found it possible to regard themselves as Christian Darwinians. In Chapter 2 R.J. Berry introduces the scientific basis of Darwinism, tracing the development of evolutionary biology from Darwin's own thought, through the neo-Darwinian synthesis of the 1930s and 1940s, to some more recent aspects of the scientific debate. He goes on to offer his own distinctive theological reading of evolutionary biology, proposing that although the scientific account of human evolution is true as far as it goes, it does not explain how humanity may be understood to be called into relation with God. This relationship required, according to Berry, a transformation of humanity from the biological species *Homo sapiens* to the species made in the image of God, *Homo divinus*. In Chapter 3 Celia Deane-Drummond welcomes Berry's outline of the development of Darwinism, particularly his setting of this in the context of Darwin's personal biography and his acknowledgment of the mixed initial reception of Darwin's ideas within both the scientific and religious communities. However, she robustly criticizes Berry's notion of a historical transformation of humanity by a special divine act not describable in terms of evolutionary biology. In doing so she argues against a literal interpretation of the accounts of the early chapters of Genesis, without denying the theological truthfulness of those accounts on matters such as the fallenness of human beings. It is possible, she suggests, to accept Darwinism and yet remain religiously conservative, and it is not necessary to assume that the two can only be held together by denying the fully natural origin of humankind.

In Chapter 4 John Hedley Brooke considers how natural theology fared in the aftermath of Darwin's *Origin of Species*. He reminds us that traditional natural theology – the deduction of God's existence and attributes from the details of nature – was already under pressure even before Darwin's theory saw the light of day. He goes on to argue that, while it is important to recognize that not all religious responses to Darwin were negative, nevertheless there was great diversity among the sympathetic responses, and that this diversity was often itself theologically divisive. In the course of this theological wrestling with Darwinism traditional natural theology diversified into various species of argument concerned mostly with consilience and based on religiously-committed reflection on nature, rather than claiming to be stand-alone arguments for theism. David Knight follows in Chapter 5 by offering further examples of the diversity of religious response to the *Origin*, suggesting that the early varieties of Christian Darwinian fall into four distinct species: parson naturalists, Christian agnostics, Incarnationalists, and pillars of the Establishment. He reminds us of advice sometimes given to historians – to read until you 'hear the voices' – and suggests that when we hear the voices of the early respondents to Darwin we can understand why it was, and still is, possible to be a Christian Darwinian.

Part 2 tightens the focus on the implications of Darwinism for natural theology. In Chapter 6 David Fergusson outlines a typology of natural theologies, arguing that each of the types (which he acknowledges may themselves be sub-dividable) has flourished under specific historical and intellectual conditions. Type 1 natural theologies claim to offer an alternative and superior route to knowledge of God compared to that afforded by revealed theology. Type 2 natural theologies are less sceptical about revealed theology, but nevertheless claim to offer a necessary rational grounding for knowledge of God derived from the testimony of scripture, signs and wonders. Type 3 natural theologies are exemplified by Thomas Aquinas, for whom the human capacity to attain some knowledge of God by means of reason is complementary to, and neither superior to or normative for God's gracious self-revelation in scripture. Interpreters differ, however, in their views of precisely how this relation is understood by Aquinas. Type 4 natural theologies are those employed for the apologetic role of "defeating the defeaters," offering arguments against the various rational criticisms of faith. In this role natural theology has "a defensive and subsidiary role, rather than a foundational or preparatory one." Finally, Type 5 natural theologies seek (merely) to establish the possibility of coherence between the specific claims of Christian faith and knowledge of the world derived from other disciplines or areas of life. Fergusson traces the effects of Darwinism on natural theology in terms of the conceptions of the respective roles of scientific and theological explanation, the relation of chance to divine providence, the intensification of the problem of evil, and the perceived threat to human significance. He suggests that the upshot, through the twentieth century, was a general abandonment of (strong) type 1 and type 2 natural theologies in favour (*pace* Barth) of recognition of a significant but auxiliary role for (weaker) type 4 and type 5 approaches.

Sarah Coakley responds in Chapter 7 by lamenting that Fergusson's analysis in effect concedes that theology must be on the back foot in relation to an epistemological agenda set by (functionally atheistic) science. She proposes instead that Fergusson's type 3 natural theology should be developed in such a way as to avoid the assumption that science, philosophy and theology must all operate within a 'flat' epistemological plane determined by science. This bolder approach, she suggests, would see theologians presenting challenges to scientists about the philosophical and metaphysical basis of current scientific research programmes, with the promise that such interventions would make concrete differences to the actual practice of experimental science. (The counterpoint of Professor Fergusson's fine analysis and Professor Coakley's response was a pivotal moment in the conference, accented by the latter's characteristically pithy delivery. I am therefore glad that Sarah agreed to allow a sense of the dynamism of that interaction to be preserved by presenting her paper for publication largely as delivered, without further scholarly embellishment.)

One possible candidate for a contemporary form of natural theology is Intelligent Design (ID) theory, which is predicated on the claim that Darwinism (and naturalistic explanations in general) cannot explain certain kinds of biological complexity. In Chapter 8 Denis Alexander delivers a thorough critique of ID theory in the form of an extended review of Stephen Meyer's recent book *The Signature in the Cell*. Alexander finds ID to be flawed scientifically (in curtailing science's search for explanations for things not currently understood but for which naturalistic explanations can reasonably be anticipated) and philosophically (in presenting ID theory as science even though it fails to meet many of the criteria by which scientific forms of enquiry may be identified). He also finds it wanting theologically, in that it fails to recognize that the Christian doctrine of creation is concerned with the absolute dependence of the created order on God, not with how particular instances of biological complexity arose. Alexander's contribution is the only chapter that was not presented in substantially its current form at the original conference. For that reason Sjoerd Bonting's response at the conference is not reproduced here, though Neil Spurway (Chapter 14) recalls a telling question of Bonting's concerning the appropriate theological interpretation of the occurrence of deleterious genetic mutations.

In Chapter 9 Christopher Southgate and I outline some of our work on the use of C.S. Peirce's semiotics (theory of signs) as a mediating philosophical framework for science and theology. On the theological side we apply this framework to the Christian understanding of God as Trinity and the idea that Jesus was the incarnate Word of the Father. On the scientific side we report how the field of semiotics my illuminate and stimulate work in fields as diverse as origin of life research and the study of human evolution. The result, we suggest, is an example of the kind of audacious version of Type 3 natural theology advocated by Sarah Coakley in Chapter 7. Kenneth Wilson responds to our paper in Chapter 10, emphasizing the significance of the communal nature of human enquiry and the way in which the Peircean perspective encourages a view of theology as a vital dimension of such enquiry.

Part 3 brings us to the contributed short papers that were presented at the conference. The first two of these explore the relation between Darwinism and natural theology in the context of the work of two theologians, one Reformed and the other Catholic, each of whose thought has been highly influential in the debate. In spite of Karl Barth's well known strictures against natural theology, in Chapter 11 Philip Chapman finds methodological reasons for pursuing the possibility of dialogue between a Barthian approach to revelation and scientific understandings of the human person. In Chapter 12 Michael Burdett argues that Pierre Teilhard de Chardin's religious appropriation of the cosmic narrative of evolution still has something to offer by way of theological reflection on the emergence of humankind through (and beyond) natural processes, even though Teilhard's biology is arguably more Bergsonian than Darwinian.

In Part 4 we are offered some wider perspectives on the opportunities and challenges presented by Darwinism to theology. In Chapter 13 Fabien Revol, in the third of the contributed papers from the conference, suggests that Darwinism offers seven gifts to theology. These are: a new framework for thinking about nature; the end of static conceptions of the universe; the possibility of a new understanding of divine design; the eviction of certain non-Christian conceptions of God; a kenotic approach to the problem of suffering; the possibility of a new understanding of the ultimate purposes of God in creation; and the recovery of the theme of divine immanence. Together these gifts point away from an old-style approach to natural theology and towards a form of theology of nature; that is, towards "[seeking] to interpret the theological meaning of nature's history rather than attempting to discover or deduce the existence of God and his attributes from the causal details of nature."

Finally, in Chapter 14, Neil Spurway offers a critical overview of the volume. This leads him to suggest four points for further reflection. First, the enrichment of the gene-focussed twentieth century approaches to Darwinian evolution by the newer multi-dimensional picture that is emerging under the banner of 'eco-evo-devo' may have fresh implications for theology. The resulting effect on theology will be, he suggests, similar in kind to that adopted by some of the earliest Christian Darwinians; namely, an ever stronger view of God's indwelling in the material order of creation. Second, he wishes to press, further than is explicit in any of the conference papers, the way in which Darwinism may point us away from certain conceptions of the nature of divine action. Indeed, he finds good theological reasons, implicit in some of the chapters, for regarding the otherness of God as so absolute that questions about divine action must always shade into mysticism and apophatic humility. Third, he wishes to emphasize the limitations of evolved minds when it comes to speaking intelligibly of that which transcends the material world to which the capacities of such minds are adapted. Fourth, and summing up the previous three points, he contends that no theology worth its salt can be anything other than a 'natural' theology.

Not surprisingly, the authors of these conference papers do not speak with a single voice. Perhaps the greatest contrast is that between the approaches of the first and the last. R.J. Berry's way of relating the theological concept of the *imago dei* to the scientific account of human evolution arguably gives theology an unusually free rein in relation to scientific knowledge. In contrast, Neil Spurway's view of the evolutionary constraints on epistemology implies a theology firmly bridled by science. Some readers may judge that one or other of these approaches – the one effectively giving the leading role to theology and the other to science – offers the only plausible way of conceiving how they might travel together. On the other hand, the reader may want to consider whether the two pivotal chapters in the book – David Fergusson's typology of natural theologies and Sarah Coakley's challenging response – provide a key to reading the other chapters as pointers to a new way of integrating the enterprises of science and theology, one which suggests a necessary (but not necessarily symmetrical) yoking of the two.

PART I:

DARWINISM AND CHRISTIAN DARWINIANS

CHAPTER TWO

BIOLOGY SINCE DARWIN

R.J. BERRY

Charles Darwin was born in Shrewsbury on 12th February 1809, the same day as Abraham Lincoln. It has been said that Lincoln liberated slaves from physical bondage while Darwin liberated thought from the constraints of out-dated views of God and nature.[1] Darwin attended the local grammar school, where he did not excel. He wrote in his *Autobiography*, "When I left the school, I was for my age neither high nor low in it; I believe that I was considered by all my masters and by my father as a very ordinary boy, rather below the common standard in intellect. To my deep mortification my father once said to me, 'You care for nothing but shooting, dogs and rat-catching; you will be a disgrace to yourself and all your family.'"[2]

The young Darwin may have been an indifferent scholar, but he was an avid collector of minerals and insects and a dedicated observer of the natural world. His response to reading Gilbert White's *Selborne* was to make notes on the habits of birds; he recorded that "In my simplicity I remember wondering why every gentleman did not become an ornithologist."[3]

The reality was that Charles Darwin was first and foremost a naturalist. Fifty years after the announcement of natural selection at the Linnean Society, Alfred Russel Wallace declared that Darwin and he had hit on the mechanism for evolution because "in early life both Darwin and myself became ardent beetle-hunters … [we] had an intense interest in the mere variety of living things."[4] Darwin's own apprehension of himself was: "I

---

[1] William E. Phipps, *Darwin's Religious Odyssey* (Harrisburg: Trinity International, 2001).

[2] Nora Barlow, ed., *The Autobiography of Charles Darwin 1809-1882* (London: Collins, 1958), 28.

[3] Barlow, *Autobiography of Charles Darwin*, note 2, 45.

[4] The Darwin-Wallace celebration held on Thursday, 1st July 1908, by the Linnean

have no great quickness of apprehension or wit which is so remarkable in some clever men, for instance Huxley. I am therefore a poor critic; a paper or book, when first read, generally excites my admiration, and it is only after considerable reflection that I perceive the weak points. My power to follow a long and purely abstract train of thought is very limited; I should never have succeeded with metaphysics or mathematics. ... On the favourable side of the balance, I think I am superior to the common run of men in noticing things which easily escape attention, and in observing them carefully. My industry has been nearly as great as it could have been in the observation and collection of facts."[5]

Six months after Darwin's birth, Jean Baptiste Lamarck published his *Philosophie Zoologique* (on 14 August 1809). It was inspired partly by theology. In it, Lamarck avoided the difficulty of extinction, which implied that God's original creation was not perfect, and suggested a progressive upward path for humankind. It is commonly portrayed as a precursor to the *Origin of Species*. This is not so.[6] Darwin had a low opinion of Lamarck's science. Although he wrote in the 'Historical Sketch' added to the Third Edition of the *Origin* that Lamarck had done the "eminent service of arousing attention to the probability of change in the organic world," he wrote to Joseph Hooker (11 January 1844), "Heaven forfend me from Lamarck['s] nonsense of a 'tendency to progression' 'adaptations from the slow willing of animals' &c.—but the conclusions I am led to are not widely different from his—though the means of change are wholly so" and again (10 November 1844) "Lamarck's is veritable rubbish."[7]

1859 and After

Darwin followed his father and elder brother to medical school in Edinburgh (1825-27), but found himself too squeamish for a medical career. He transferred to Cambridge, reading for a general degree (1828-31). This was followed by five much more exciting years (1832-36) as a "gentleman naturalist" on *HMS Beagle*, commissioned under the command of Robert Fitzroy to survey the southern coasts of South America. The

---

Society of London (London: Linnean Society, 1908), 7.

[5] Barlow, *Autobiography of Charles Darwin*, note 2, 140.

[6] Although in Britain they were welcomed by intellectual radicals as a justification for attacking the social status quo. Adrian Desmond, *The Politics of Evolution* (Chicago: Chicago University Press, 1989).

[7] Frederick Burkhardt and Sydney Smith, *The Correspondence of Charles Darwin, Volume 4 1844-1846* (Cambridge: Cambridge University Press, 1988), 1 and 77.

Beagle went on to circumnavigate the globe, most famously spending five weeks (16 September to 20 October 1835) around the islands of the equatorial Galapagos Archipelago, 1000 km west of the South American mainland.

Darwin never left Britain again. His scientific reputation was made by his account of the *Voyage of the Beagle*, published in 1839. In the same year he was elected a Fellow of the Royal Society and married his cousin, Emma Wedgwood (granddaughter of the founder of the Wedgwood pottery firm). The couple lived first in London (in a house where the Biology Department of University College London now stands) and then at Down House near Bromley in Kent until their deaths – Charles in 1882 and Emma in 1896.

Darwin's assumption had been that he would seek ordination after his Cambridge degree. After leaving Edinburgh, he had read with approval the evangelical Bishop of Chester's *Evidences of Christianity.*[8] At Cambridge, he was required to study William Paley's *Evidences of Christianity;*[9] he found Paley's logic "irresistible." In his *Autobiography* he notes, "The logic of this book and I may add of his *Natural Theology* gave me as much delight as did Euclid. The careful study of these works … was the only part of the Academic Course which, as I then felt and still believe, was of the least use to me in the education of my mind."[10] He wrote home at an early stage of the *Beagle*'s voyage, "Although I like this knocking about, I find I steadily have a distant prospect of a very quiet parsonage & I can see it even through a grove of Palms."

But during his time on the *Beagle*, he began to drift away from the idea of career as a clergyman. He did not become an atheist; in his *Autobiography* he insisted that he continued to believe in some form of God. Three years before his death he wrote to a correspondent, "I have never been an atheist in the sense of denying the existence of God." Notwithstanding, as Janet Browne comments,

> It is clear that his kind of belief, though orthodox, was a very loose, English-style orthodoxy in which it was far less trouble to believe than it was to disbelieve. … For Darwin, as for countless others, belonging to the Church of England was as much a statement of social position and attitude as it was a profession of any particular doctrine. … No sane man could

[8] J.B. Sumner, *Evidences of Christianity derived from its Nature and Reception* (London: J. Hatchard and son, 1821). Sumner was Archbishop of Canterbury 1848-62.

[9] William Paley, *Evidences of Christianity* (London, 1802).

[10] Barlow, *Autobiography of Charles Darwin*, note 2, 59.

believe in miracles, he decided. … Yet he went to church regularly throughout the voyage, attending the shipboard ceremonies conducted by Fitzroy and services on shore whenever possible."[11]

Paley's comfortable picture of a contented creation came increasingly under pressure during the first half of the nineteenth century through:

- The discovery of 'deep space' as telescopes improved. In 1783 William Herschel had shown that the Milky Way itself was moving through space.
- The acceptance from fossils and geological strata that the universe was much older than the few thousand years implied by the biblical genealogies. Such 'deep time' was the Achilles heel of traditional natural theology. A creator could presumably design an organism perfectly adapted to a particular environment, but this perfection would disappear if the environment was not constant. Adaptation to changes in climate, to the physical structure of the Earth's surface, or to predators and competitors is possible only if organisms change. By Darwin's time, the majority of Bible expositors accepted a non-literal interpretation of the early chapters of Genesis.[12]
- The increasing knowledge of the distribution of animals and plants from scientific voyaging – by Banks, Solander and Herschel, by Humboldt, Darwin, Hooker and others. Noah's Ark faded into the history of thought;[13] the possibility of multiple centres of creation became increasingly discussed.

Darwin was, of course, aware of these tensions, and found the Platonic stasis of Paley increasingly difficult as he read the ideas of gradual geological change propounded by Charles Lyell in his *Principles of Geology*.[14] During his time on the *Beagle*, he was faced with:

---

[11] Janet Browne, *Charles Darwin: Voyaging* (London: Jonathan Cape, 1995), 325.
[12] C.L.E. Lewis and S.J. Knell, eds., *The Age of the Earth: from 4004 BC to AD 2002* (London: Geological Society Special Publication No. 190, 2001).
[13] Janet Browne, *The Secular Ark: Studies in the History of Biogeography* (New Haven: Yale University Press, 1983).
[14] Darwin took the first volume of the *Principles* with him when he embarked on the Beagle; he received the second volume in Montevideo; and the third volume when the ship reached the Falkland Islands.

- South American fossils similar but obviously different to living forms. In the Argentine he was astonished to find a fossil of a giant sloth (*Megatherium*) as big as an elephant.
- Replacement of species with latitude, apparent as the *Beagle* sailed south along the South American coast.
- Major effects on topography and earthquakes affecting land uplift; Darwin experienced such an event in Chile.
- The occurrence of many unusual species in archipelagos.

It is now common knowledge that Darwin did not have a Damascus Road experience in the Galapagos as was once believed.[15] Notwithstanding, he recorded his belief in the possible importance of islands in his notes made after leaving the Galapagos.

The *Beagle* arrived back in Britain in October 1836. Six months later (March 1837), Darwin was told by John Gould, the ornithologist at London Zoo that the Galapagos Islands were home to a number of different bird species; in September of the same year he read "for amusement" *An Essay on the Principle of Population* by Thomas Malthus. His ideas of a mechanism for evolutionary change (or 'transformism' in the language of the time) began to gel. His starting point was a Paleyian belief that the creator creates by laws.[16] By 1842, he was sufficiently sure of his new understanding to describe it in a brief 35 page 'Sketch'; he expanded this into a 200 page 'Essay' in 1844. The latter subsequently formed the basis of the *Origin of Species*, published in 1859.

Darwin invoked a very simple mechanism for evolutionary change, based on three facts and two deductions from these facts. He started with the observation that virtually all species have a large potential for increase in number (think of the number of acorns produced by an oak tree or the masses of spawn laid by every female frog), but (second observation) numbers remain roughly constant. The inference from this is that there must be a *struggle for existence*, with only a small proportion of young surviving. The existence of such a struggle is essentially an ecological deduction and one well understood in Darwin's time; it had forced itself on Darwin's awareness when he read Malthus's *Essay*. Darwin's genius was in linking a third fact – heritable variation – to the ever-prevalent

---

[15] Frank J. Sulloway, "Darwin and his finches. The evolution of a legend," *Journal of the History of Biology* 15: 1-53.
[16] Dov Ospovat, *The Development of Darwin's Theory* (Cambridge: Cambridge University Press, 1979).

struggle for existence.[17] If only a small proportion of a population survive the struggle, the likelihood is that they will include a high proportion of those with any trait which gives them some sort of advantage. Over the generations, those carrying the trait will increase at the expense of those lacking the trait. This would lead to a genetic change in the population, amounting to *natural selection* for the trait in question.

The *Origin of Species* was published on 24 November 1859, price 15 shillings (£0.75). The reason for the delay in publishing has been the subject of much speculation. One factor seems to have been Darwin's consideration for his wife's views. Emma was a quiet but devout Anglican. A month before their marriage she had written to Charles:

> My reason tells me that honest and conscientious doubts cannot be a sin, but I feel it could be a painful void between us. I thank you from my heart for your openness with me & I should dread the feeling that you were concealing your opinions from the fear of giving me pain. It is perhaps foolish of me to say this much but my own dear Charley we now do belong to each other & I cannot help being open with you. Will you do me a favour? yes I am sure you will, it is to read our Saviours farewell discourse to his disciples which begins at the end of the 13th Chap of John. It is so full of love to them & devotion & every beautiful feeling. It is the part of the New Testament I love best. This is a whim of mine it would give me great pleasure.

There were at least two other – scientific – reasons for Darwin's procrastination. He was seriously alarmed in 1844 by the uproar provoked by the publication of *Vestiges of the Natural History of Creation* (published anonymously, but actually written by Robert Chambers), effectively a tract against the deism of William Paley's version of natural theology.[18] Chambers wrote that when there is a choice between special creation and the operation of general laws instituted by the creator, "I

---

[17] In his autobiography, Darwin wrote: "At that time I overlooked one problem of great importance … the tendency in organic beings descended from the same stock to diverge in character as they became modified. That they diverged greatly is obvious from the manner in which species of all kinds can be classed under genera, genera under families, families under sub-orders, and so forth; and I can remember the very spot in the road, whilst in my carriage, when to my joy the solution came to me; and this was long after I had come to Down [1842]. The solution, as I believe, is that the modified offspring of all dominant and increasing forms tend to become adapted to many and highly diversified places in the economy of nature" (Barlow, *Autobiography of Charles Darwin*, note 2, 120-1).

[18] James A. Secord, *Victorian Sensation* (Chicago: University of Chicago Press, 2000).

would say that the latter is generally preferable as it implies a far grander view of the divine power than the other." Since there was nothing in the inorganic world "which may not be accounted for by the agency of the ordinary forces of nature," why not consider "the possibility of plants and animals having likewise been produced in a natural way."[19]

The *Vestiges* was an immediate best-seller. In the ten years following its publication, it sold more copies than did the *Origin* fifteen years later. But it was full of errors. For Darwin, "the writing and arrangement are certainly admirable, but the geology strikes me as bad & his zoology far worse."[20] Others roundly condemned the book. The geologist Adam Sedgwick lambasted it in an 85 page diatribe in the *Edinburgh Review*; he called it "a foul book [in which] gross credulity and rank infidelity join in unlawful marriage."[21] He wrote to his colleague Charles Lyell, "If the book be true, the labours of sober induction are in vain; religion is a lie; human law is a mass of folly and a base injustice; morality is moonshine; our labours for the black people of Africa were works of madmen; and man and woman are only better beasts." Its importance is the debate it stirred. Despite his misgivings about it, Darwin welcomed the book on the grounds that "it has done excellent service in calling in this country attention to the subject and in removing prejudices."

Not long after the *Vestiges* appeared, Joseph Hooker, Darwin's friend and confidant, commented in a letter to Darwin (September 1845) (*a propos* a book on the nature of species by a French botanist, Frédéric Gérard, whom Hooker had disparaged because he was "neither a specific naturalist, nor a collector, not a traveller … and therefore a distorter of facts"), that to be qualified to speculate about the nature of species, "one must have handled hundreds of species with a view to distinguishing them & that over a great part – or brought from many parts – of the globe. I am not inclined to take much for granted from any one who treats the subject in his way and who does not know what it is to be a specific Naturalist himself." Hooker seemed to assume that Darwin fulfilled these requirements, but Darwin was alarmed. His knowledge of the natural world was extensive, but it was not deep. He responded to Hooker, "No one has the right to examine the question of species who has not minutely

[19] Robert Chambers, *Vestiges of the Natural History of Creation* (London: John Churchill, 1844).

[20] Letter to Joseph Hooker, 7 January 1845 (letter no. 814 in the Darwin Correspondence Project Archive, Cambridge University Library).

[21] *Edinburgh Review*, July 1845. Sedgwick confessed that "reviewing was more like writing a sermon than composing a scientific paper" (Secord, *Victorian Sensation*, note 18, 246).

examined many."[22] He began a study of the barnacles he had collected on the *Beagle* cruise. This stretched into a detailed investigation of barnacles worldwide and occupied him from 1846 to 1854, culminating in four massive monographs, two on living barnacles and two on fossil forms, an achievement which still stands as a definitive treatment of the group.[23] The Royal Society gave him a Royal Medal for the work. But Darwin's barnacle explorations were more than a self-justifying exercise: they opened his eyes to variation within species and the apparent relationships of different families. When he returned to his 'transformism' he was much more aware than previously of the extent of biological complexity in the real world.[24]

When the *Origin* eventually appeared, it was anathematized by some, but welcomed by many others. One of the first responses to Darwin came in a letter from Charles Kingsley, at the time Rector of Eversley in Hampshire but soon to become Regius Professor of Modern History at Cambridge. Kingsley wrote, "I find it just as noble a conception of Deity to believe that He created primal forms capable of self development … as to believe that He required a fresh act of intervention to supply the *lacunas* which He himself had made."[25] Evangelicals were more accepting than liberals, apparently due to their stronger doctrine of God's providence ('common grace').[26] Ironically in view of later history, a number of the authors of the 'Fundamentals', booklets produced between 1910 and 1915 to expound the 'fundamental beliefs' of Protestant theology as defined by the General Assembly of the American Presbyterian Church and which have led to the word 'fundamentalism' entering the language, were sympathetic to evolution. Princeton theologian B.B. Warfield, a passionate advocate of the inerrancy of the Bible, argued that evolution could provide a tenable "theory of the method of divine providence in the creation of mankind."[27]

---

[22] Letter to Joseph Hooker, 10 September 1845 (letter no. 915 in the Darwin Correspondence Project Archive, Cambridge University Library).

[23] Rebecca Stott, *Darwin and the Barnacle* (London: Faber & Faber, 2003).

[24] Richard Bellon, "Joseph Hooker takes a 'fixed post': transmutation and the 'present unsatisfactory state of systematic botany', 1844-1860," *Journal of the History of Biology* 39 (2006): 1-39.

[25] Letter to Charles Darwin, 18 November 1859 (letter no. 2534 in the Darwin Correspondence Project Archive, Cambridge University Library).

[26] David N. Livingstone, *Darwin's Forgotten Defenders: The Encounter Between Evangelical Theology and Evolutionary Thought* (Grand Rapids: Eerdmans, 1987).

[27] B.B. Warfield, "On the antiquity and unity of the human race," *Princeton Theological Review* 9 (1911): 1-25. Cited in James R. Moore, *The Post-Darwinian Controversies* (Cambridge: Cambridge University Press, 1979), 71.

A major early objection to the *Origin* was Darwin's use of deduction rather than the then accepted discipline of Baconian induction.[28] This was a case of the biter bit. Soon after his return from the *Beagle*, Darwin wrote about the prominent 'parallel roads' along the sides of Glen Roy near Fort William. Drawing on his observations in South America and using strict inductive arguments, he claimed they indicated the margins of ancient seas, raised to their present position by uplift of the mountains surrounding the glen. He was wrong: the 'roads' were soon shown to be the margins of glacial lakes formed during the Pleistocene. Darwin attempted to defend his interpretation, but had eventually to concede that he had made a "gigantic blunder."[29] The so-called hypothetico-deductive method to which Darwin turned and which he used in the *Origin* is nowadays virtually universally accepted as the proper way of doing science.

The mass of evidence for the fact of evolution led to a rapid acceptance of 'transformism'. The infamous debate between the Bishop of Oxford and Thomas Huxley at the 1860 meeting of the British Association for the Advancement of Science was not really about evolution *versus* creation or even science *versus* religion. On the Bishop's side it was about the danger of legitimising change in an age when he believed it was having dangerous social and theological effects; Huxley's aim was the secularisation of society – establishing the legitimacy of science against what he regarded as the improper influence of church leaders.[30] It was reported that Wilberforce went away happy that he had given Huxley a bloody nose, while Joseph Hooker (who spoke after Huxley) told Darwin that Huxley had been largely inaudible. As far as the audience was concerned, many scored it as an entertaining draw.[31] The tragedy has been a legacy of

---

[28] Moore, *The Post-Darwinian Controversies*, 194ff.

[29] Martin J.S. Rudwick, "Darwin and Glen Roy: a 'great failure' of scientific method," *Studies in History and Philosophy of Science* 5 (1974): 97-185.

[30] The other well-known evolution-creation contest – the 1926 Scopes "Monkey Trial" in Dayton, Tennessee – was also really about non-science issues: the self-promotion of a small town, the freedom of expression under the American constitution, the agenda of a populist politician (William Jennings Bryan), and the ambitions of the American Civil Liberties Union for a test case (Edward J. Larson, *Summer for the Gods: The Scopes Trial and America's Continuing Debate over Science and Religion*. Cambridge: Harvard University Press, 1997; Karl W. Giberson, *Saving Darwin*. New York: HarperCollins, 2008). Like the Wilberforce-Huxley debate, and helped by a Broadway play later filmed as *Inherit the Wind*, it has become a formidable myth about the triumph of science over prejudiced religion.

[31] John Hedley Brooke, "The Wilberforce-Huxley debate: why did it happen?" *Science & Christian Belief* 13 (2001): 127-141.

inexorable conflict between science and faith, encouraged by Huxley himself and fuelled by two much-read manifestos by John William Draper (*History of the Conflict between Religion and Science,* 1875) and Andrew Dickson White (*A History of the Warfare of Science with Theology in Christendom*, 1886).[32]

It is worth noting that Darwin (like Wallace) was primarily concerned to describe a possible *mechanism* for evolutionary change, not with the actual occurrence of change. The latter was rapidly accepted by the majority because it made sense of so many diverse facts. In contrast, Darwin's 'big idea' of natural selection was rather unsuccessful; it did not fit easily with conventional notions of progress or improvement, and philosophers and theologians developed alternative proposals – more metaphysical than scientific.[33]

Even worse, the rediscovery of Gregor Mendel's results in 1900 and the subsequent explosion of the science of genetics seemed to show that the physical basis of heredity could not be the basis of the gradualism expected by Darwinism. Mutations studied in the laboratory tended to have a large effect, to be deleterious, and to be inherited as recessive traits. The Darwinian expectation was that evolution progressed through small steps produced by favourable variants. Surveying the state of evolutionary biology for the Jubilee of the *Origin* in 1908, Vernon Kellogg spoke of "the death-bed of Darwinism." He wrote, "Darwinism as the all-sufficient or even the most important causo-mechanical factor in species-forming

---

[32] Colin Russell comments: "Today the historical views of Draper and White are totally unacceptable, not merely because of many factual aberrations, but much more because they represent a long demolished tradition of positivist, Whiggish historiography. … Draper … was gripped by the fear of the power wielded by the Roman Catholic church. … White was President of Cornell, the first non-sectarian university in the USA. His enemies had been those advocates of sectarian theology to whom the very existence of Cornell was a tangible threat. His book was directed not so much against religion as against dogmatic theology" (Colin Russell, "The conflict metaphor and its social origins," *Science & Christian Belief* 1 (1989): 6).

[33] It is worth noting that the mechanism (often called 'Darwinism') is manifestly open to test and therefore possible of disproof, contrary to the claim often made by modern anti-evolutionists that it is dogma rather than science. The widely-quoted criticism (albeit later retracted) by Karl Popper that "Darwinism is not a testable scientific theory but a metaphysical research programme" was based on his rejection of all 'historical' sciences (because they were non-falsifiable), plus a misunderstanding of the Darwinian mechanism. Popper made his retraction in a letter to the *New Scientist*, 87 (21 August 1980): 611. For a full account of this episode, see David N. Stamos, "Popper, falsifiability and evolutionary biology," *Biology and Philosophy* 11 (1996): 161-191.

and hence as the sufficient explanation of descent, is discredited and cast down."[34] Into this apparent void, an extravagance of other evolutionary theories poured: Berg's *Nomogenesis*, Willis's *Age and Area*, Smuts's *Holism*, Driesch's entelechy, Osborn's aristogenesis and orthogenesis. Invention was rife. A common feature was some form of inner progressionist urge or *élan vital.* Unfortunately three standard and still-read histories of biology (by Nordenskiöld, Rádl and Singer) were written during this time, perpetuating the idea that evolutionary theory is an illogical mess and that Darwinism is completely eclipsed.

The scientific confusion spread into theology. The idea that evolution was driven by some sort of purpose was initially espoused by some distinguished scientists – the zoologist Ray Lankester and the physiologist J.S Haldane, the psychologists Lloyd Morgan, William McDougall and E.S. Russell, physicists like Oliver Lodge, and the cosmologists Arthur Eddington and James Jeans; as well as by popularisers like Arthur Thomson and politicians such as Arthur Balfour. Not surprisingly with such apparently informed authorities, the idea of directed evolution was seized upon by churchmen, prominent among them being Charles Gore, and somewhat later by W.R. Inge, Hensley Henson, Charles Raven and E.W. Barnes. This cross-over of evolutionary idealism from science to theology has been elegantly chronicled by Peter Bowler.[35] It eventually died through its perceived ineffectiveness rather than conscious rejection. "The Modernists saw themselves marginalized not by the new science, of which many remained unaware, but by changing values within the churches, which brought back a sense of human sinfulness and alienation from God incompatible with the idea of progress".[36] One can have some sympathy with the theologians. It took the scientists a long time to reach an evolutionary synthesis, but this does not excuse the uncritical use of inadequate science.

The Neo-Darwinian Synthesis

The irrelevance of the evolutionary speculation provoked by the early Mendelians was exposed by a series of theoretical analyses in the 1920s, beginning with two difficult and largely non-understood papers by R.A. Fisher, in which he showed that continuous variation could arise through

---

[34] Vernon L. Kellogg, *Darwinism Today* (London: George Bell & Sons, 1907), 374.

[35] Peter J. Bowler, *Reconciling Science and Religion* (Chicago: Chicago University Press, 2001).

[36] Bowler, *Reconciling Science and Religion*, note 30, 417.

the cumulative effect of many genes, each with a small effect, and that dominance was the result of interaction between genes rather than an intrinsic property of a gene by itself.[37] He argued that the dominance or recessivity of any character is the consequence of repeated mutation during evolutionary time, with modification of its effect by other genes – towards greater expression if its effects are beneficial to its carrier or lesser if it has deleterious effects; *i.e.*, its expression will be modified towards either dominance or recessivity, thus removing the difficulties raised by the early Mendelians.

Fisher's analyses were complemented by J.B.S. Haldane in Britain and Sewall Wright in the United States, and summarized in a series of major works.[38] They were supported by studies of inherited variation in natural populations by E.B. Ford in Britain and Theodosius Dobzhansky in the United States.[39] Their conclusions, together with results from many other sources, were brought together by Julian Huxley in a book *Evolution: the Modern Synthesis* (1942), which provided the eponymous name to the incorporation of Mendelian genetics into the insights of Darwin, and the final reconciliation of the earlier evolutionary debates.[40]

The Mendelian or neo-Darwinian synthesis of the 1940s has furnished a robust understanding of evolutionary processes as well as a justification of Darwin's original theory.[41] Its most serious challenge came in the 1960s when biochemical techniques of protein analysis were applied to variation in natural populations and unexpectedly large amounts of inherited variation were discovered – between one in four and one in twenty genes tended to be represented by slightly different forms (*alleles*) on the chromosomes an individual receives from his or her two parents.

---

[37] R.A. Fisher, "The correlation between relatives on the supposition of Mendelian inheritance," *Transactions of the Royal Society of Edinburgh* 52 (1918): 399-433; R.A. Fisher, "On the dominance ratio," *Proceedings of the Royal Society of Edinburgh* 42 (1922): 321-41.

[38] R.A. Fisher, *The Genetical Theory of Natural Selection* (Oxford: Clarendon Press, 1930); Sewall Wright, "Evolution in Mendelian populations," *Genetics* 16 (1931): 97-159; J.B.S. Haldane, *The Causes of Evolution* (London: Longmans, 1932).

[39] E.B. Ford, *Mendelism and Evolution* (London: Methuen, 1931); Theodosius Dobzhansky, *Genetics and the Origin of Species* (New York: Columbia University Press, 1937).

[40] Julian Huxley, *Evolution: The Modern Synthesis* (London: Allen & Unwin, 1942).

[41] Ernst Mayr and William B. Provine, eds., *The Evolutionary Synthesis* (Cambridge: Harvard University Press, 1980); R.J. Berry, *Neo-Darwinism* (London: Edward Arnold, 1982).

Conventional understanding was that such high levels of variation would lead to an insupportable 'genetic load' because the less advantageous allele would reduce the reproductive potential of its carrier. The simplest escape from this dilemma was to assume that such biochemical variants had no effect on their bearer; *i.e.*, that they were 'neutral' and thus not subject to selection. This argument seemed to be supported by apparently regular rates of accumulation of new variants (mutations), to the extent that a 'molecular (or protein) clock' could be calibrated on the basis of the number of gene differences between two lineages.

However it soon appeared that the protein clock did not keep good time. Different proteins could change at rates differing by two orders of magnitude while the same protein may change faster (or slower) in different groups. These are effects that would be expected if the proteins were subject to selection rather than a physically determined mutation rate. The falsifying of extreme neutralist assumptions led to attention being re-focussed on environmental (or ecological) factors in evolution. For example, it is improper to speak of *the* selective effect of an inherited character. Selection varies in both time and space. It may be density or frequency dependent or independent. In the well known example of moths being selectively predated by birds, the chance of being eaten (*i.e.* the intensity of selection) varies with the amount and history of atmospheric pollution. Places where black moths had a high survival rate have changed as pollution has declined following Clean Air legislation, and the survival rate of black moths has declined in proportion.

A positive outcome of the neutralism debates has been to rescue evolutionary studies from the danger of over-dependence on theoretical models and lead them back to an observation and experimental basis – which is where Darwin and Wallace began.[42] A longer standing worry has been more resistant, although it is related. This is the concern that mutations occur at 'random' and hence that evolution is at the mercy of purposeless physico-chemical accidents, implying that we are ruled by biochemical accidents. The most eloquent advocate of this position has been the French Nobel Laureate Jacques Monod in his book, *Chance and Necessity*.[43] Stephen Gould has similarly argued that "rewinding the tape

---

[42] R.J. Berry, "Ecology: where genes and geography meet," *Journal of Animal Ecology* 58 (1989): 733-59; R.J Berry, T.J.S. Crawford, and G.M. Hewitt, eds., *Genes in Ecology* (Oxford: Blackwell Scientific, 1992).

[43] Jacques Monod, *Chance and Necessity* (London: Collins, 1971); originally published in Paris by Éditions du Seuil, Paris as *Le Hasard et la Necessité*, 1970. See response by W.H. Thorpe, *Purpose in a World of Chance* (Oxford: Oxford University Press, 1978).

of life" will always produce a different result; he has written of "the awesome improbability of human evolution."[44]

Many authors have argued that this understanding of randomness is contraindicated by the universal existence of adaptation to the environment.[45] This fact has been given new and convincing weight by Simon Conway Morris, who points to the commonness of 'convergence' in evolutionary lineages.[46] His thesis is that there are a limited number of solutions to any adaptive challenge, shown by the way in which similar structures and habits have repeatedly evolved independently. He records, for example, that eye lenses have appeared at least 15 times in different groups. Genes are repeatedly co-opted for new functions. He believes that evolution is "eerily predictable," wholly contradicting the accusation that it is governed by chance. As a result, the constraints of evolution and the ubiquity of convergence "strongly suggest humans to be an evolutionary inevitability."[47] These conclusions remain contentious, but they could be true and certainly show that the Gould's assumption is insecure.[48]

Only an Ape?

The biggest theological challenges for Darwinian ideas concern human evolution. In an effort to minimise controversy, Darwin steered clear of human evolution in the *Origin*. He included only one mention of the subject: "I see open fields for far more important researches. ... Much light will be thrown on the origin of man and his history."[49] However, he could not ignore the topic altogether, and returned to it fourteen years later

---

[44] Gould has used the analogy repeatedly. He is most quoted on the subject in *Wonderful Life* (New York: Norton, 1989), 45-52.

[45] For example, C.A. Coulson, *Science and the Idea of God* (Cambridge: Cambridge University Press, 1958); W.H. Thorpe, *Purpose in a World of Chance*; Donald M. MacKay, *Science, Chance and Providence* (Oxford: Oxford University Press, 1978); David J. Bartholomew, *God of Chance* (London: SCM, 1984).

[46] Simon Conway Morris, *Life's Solution. Inevitable Humans in a Lonely Universe* (Cambridge: Cambridge University Press, 2003).

[47] Simon Conway Morris, "The Boyle Lecture 2005: Darwin's compass," *Science & Christian Belief* 18 (2006): 14.

[48] Simon Conway Morris, ed., *The Deep Structure of Biology. Is Convergence Sufficiently Ubiquitous to Give a Directional Signal?* (West Conshohocken, PA: Templeton Foundation Press, 2008).

[49] In 1857, Darwin wrote to Wallace about his planned book on 'transformism', "You ask whether I shall discuss 'man'. I think I shall avoid the whole subject, as so surrounded with prejudices; though I fully admit it is the highest and most interesting problem for the naturalist."

with the *Descent of Man*. He "warned friends that it would seem 'very wicked;' Churchmen would think him 'an outcast & a reprobate;' he expected 'universal disapprobation, if not execution' – meaning, he told a critic, that the book would 'quite kill me in your good estimation'."[50]

Darwin knew almost nothing about human fossils. The first Neanderthal fossils were only described in 1857. But since that time many putative hominoid and hominid fossils have been discovered, to the extent that it is fair to claim that *Homo* has a better fossil record than almost any other genus. Unfortunately its credibility has been marred in popular understanding by over-imaginative reconstructions, notably the Piltdown debacle but also by many fanciful attempts to portray human ancestors as either hulking brutes or mere variants of modern individuals.[51] The image of human fossil history for many is probably the much-reproduced frontispiece of T.H. Huxley's *Man's Place in Nature* (1863), showing a parade of modern skeletons from a gibbon, through a series of stooping apes, to an upright man[52] ("a grim and grotesque procession" as Darwin's critic, the 8th Duke of Argyll called it), misleadingly implying evolution through living primates to modern humans. Notwithstanding, and recognising many uncertainties, there is now general agreement among specialists that the human and chimpanzee lines separated about five million years ago, and that *Homo* originated from an *Australopithecus* stock in Africa, with the first fossils classified in the genus *Homo* (*H. habilis*) occurring about 2.4 million years ago. About 1.8 million years ago a new form of *Homo* appeared in eastern Africa, *H. erectus*. It persisted in Africa for more than a million years, but also spread into Asia.[53] It had a brain size (or more strictly, cranial volume) of 800-900 ml, about a third greater than *H. habilis,* and a less projecting face. Four hundred thousand years ago, the earliest fossils regarded as *H. sapiens* (brain *c*. 1100-1300 ml) occur. They have been found in Africa, Europe and China. 'Neanderthal man' appeared around 200,000 years ago in Europe and South West Asia. It is now regarded as at least sub-specifically different

---

[50] James Moore and Adrian Desmond, introduction to *The Descent of Man*, by Charles Darwin (London: Penguin Classics, 2004), xv.

[51] A good account of this over-enthusiasm is given by Peter J. Bowler, *Theories of Human Evolution: a Century of Debates 1844-1944* (Oxford: Basil Blackwell, 1986).

[52] Thomas H. Huxley, *Man's Place in Nature* (London: Williams & Norgate, 1863).

[53] Clive Finlayson, "Biogeography and evolution of the genus Homo," *Trends in Ecology and Evolution* 20 (2005): 457-63.

from *H. sapiens*, and coexisted with the latter until disappearing 30,000 to 40,000 years ago.[54]

Developments in molecular biology have driven the recognition of our genetic closeness to the other primates. An earlier calculation that humans and chimps share 98.4% of their genes has been refined now that both genomes have been sequenced; nevertheless there is a clear identity between more than 95 per cent of the genes in the two species.[55] Notwithstanding, our DNA differences from the chimpanzees amounts to around 10,000 nucleotide changes, most of them in so-called 'junk DNA' which does not code for the production of particular proteins but regulates the activity of protein coding sequences. We have one less pair of chromosomes than all other apes (23 pairs instead of 24), but the difference is the simple result of end-to-end fusion between two separate elements of the ape chromosome set.

There do not seem to be many novel genes in the human genome. Around one in twelve of our genes apparently consists of retroviral inserts, influencing gene regulation in significant ways. One of the most intriguing results so far from comparative molecular genetics is that since the human and chimpanzee lines separated, a third more chimpanzee genes show signs of selection than do human ones. Put another way, this implies that chimpanzees are more specialized than humans; we are generalists, they are adapted to a particular niche. There is no scientific support for the notion that we have been propelled towards a predetermined end by a Divine Watchmaker, or even a 'blind' one.[56]

This is a non-intuitive result. However, many of the characteristic human features (brain size, hairlessness, prolonged adolescence) can be attributed to a neotenous change in the human line,[57] which could be genetically simple – perhaps the result of a single mutation. Furthermore, the complexity of language is a key human characteristic, without denying or denigrating the sophistication of communication in many non-human

---

[54] Chris Stringer, "Modern human origins – progress and prospects," *Philosophical Transactions of the Royal Society of London*, B, 357 (2002): 563-79; J.H. Relethford, "Genetic evidence and the modern human origins debate," *Heredity* 100 (2008): 555-63; Jonathan Silvertown, ed., *99% Ape* (London: Natural History Museum, 1998).

[55] Roy J. Britten, "Divergence between samples of chimpanzee and human DNA sequences is 5%, counting indels," *Proceedings of the National Academy of Sciences of the USA* 99(2002): 13633-35.

[56] Cf. Richard Dawkins, *The Blind Watchmaker* (London: Longman, 1986).

[57] The retention of juvenile features into adult life as a result of a slowing of the rate of maturation.

groups. The differentiating feature between ape and human is not the ability simply to make sounds, but to control those sounds precisely. "The missing ingredient [which prevented the chimps developing more complex speech] may have been a tiny change in anatomy of the proto-human vocal tract to give us finer control and permit formation of a much greater variety of sounds."[58]

Alfred Russel Wallace argued that natural selection had acted in the earlier stages of human differentiation from the apes, but as our intellectual and moral faculties became "fairly developed," the body ceased to be subject to selection, and subsequently adaptation was solely "through the action of the mind."[59] Wallace's starting point was a belief that brain size was a reliable indicator of mental capacities. The difficulty as he saw it was that both fossil humans and 'savages' (Wallace's word) had skulls (and therefore brains) of similar size to those of civilised people, and consequently all must be presumed to have the same mental capacities. However it seemed to Wallace that such traits as mathematical ability and the ability to carry out complex trains of abstract reasoning would have been useless (if not harmful) in the struggle for existence in primitive cultures. As it was both unneeded and unused, the brain could not have evolved by natural selection alone. Consequently, and certainly influenced by his belief in spiritualism, he proposed that a 'Higher Intelligence' had guided human evolution in "definite direction and for special ends."

Darwin disagreed with Wallace's conclusion that "natural selection could not have done it all," but he was himself unsure how selection might have acted to produce morality. He wrote in the *Descent of Man*, "He who was ready to sacrifice his life, as many a savage has been, rather than betray his colleagues, would often leave no offspring to inherit his noble nature".[60] A possible answer to this problem would be if a group rather than an individual is the target of selection, and there have been repeated proposals for 'group selection'.[61] However, this seems highly improbable from all that we know about biological fitness and gene transmission.

---

[58] Jared Diamond, *The Rise and Fall of the Third Chimpanzee* (London: Radius, 1991), 47.

[59] Alfred Russel Wallace, "The origin of human races and the antiquity of man deduced from the theory of natural selection," *Journal of the Anthropological Society of London* 2 (1864): 157-187.

[60] Charles Darwin, *The Decent of Man and Selection in Relation to Sex* (London: John Murray, 1871), 200.

[61] For example: V.C. Wynne-Edwards, *Animal Dispersion in Relation to Social Behaviour* (Edinburgh: Oliver & Boyd, 1962); David Sloan Wilson, *Darwin's*

It was left to J.B.S. Haldane to show a way forward, pointing out that if individual unselfishness (even to the extent of self-sacrifice) has an inherited basis so that altruistic individuals help near relatives, then 'altruistic genes' could be selected and spread within families.[62] In this way, there could be situations where cooperation (or unselfishness) is an advantage to a group of individuals, even if particular individuals are disadvantaged. Haldane's argument was formalised in 1964 by W.D. Hamilton as 'inclusive fitness', often called nowadays 'kin selection', and popularised by E.O. Wilson as a key component of 'sociobiology'.[63] Sociobiological ideas have been extremely important in biology, and have stimulated an immense amount of research. They have also provoked much dissent, particularly as they apply to mammals (especially humankind) because of the implications that behavioural choices are programmed (or 'determined', *sic*) by genes. Moreover, human social organization extends far beyond family relationships, and most observers believe that the sociobiological mechanisms so far investigated are insufficient to account fully for the widespread acknowledgement of the Golden Rule (loving ones' neighbours as oneself).

Our ancestry has certainly been shaped by selection. Are we no more than survival machines controlled by selfish genes? Will we fully understand human nature once we bring together embryology, anatomy,

---

Cathedral: Evolution, Religion and the Nature of Society (Chicago: Chicago University Press, 2002); Lee Alan Dugatkin, *The Altruism Equation* (Princeton: Princeton University Press, 2006).

62 He first put forward this idea in his 1932 book (Haldane, *Causes of Evolution*, note 33), but presented it in a more popular form in an article in 1955. "Let us suppose that you carry a rare gene which affects your behaviour so that you jump into a flooded river and save a child, but you have one chance in ten of being killed, while I do not possess the gene and stand on the bank and watch the child drown. If the child is your own child or your brother or sister, there is an even chance that the child will also have this gene, so five genes will be saved in children for one lost in an adult. If you save a grandchild or nephew, the advantage is only two and a half to one. If you only save a first cousin, the effect is very slight. If you try to save your first cousin once removed the population is more likely to lose this valuable gene than to gain it. …. It is clear that genes making for conduct of this kind would only have a chance of spreading in rather small populations where most of the children were fairly near relatives of the man who risked his life" (J.B.S. Haldane, "Population Genetics," *New Biology* 18(1955): 44).

63 W.D. Hamilton, "The genetical evolution of social behaviour," *Journal of Theoretical Biology* 7 (1964): 1-52; Edward O. Wilson, *Sociobiology* (Cambridge: Harvard University Press, 1975).

genetics, ecology and behavioural studies? Certainly such a synthesis would be a useful step, but it seems over-optimistic to expect this to give us all the answers we want. Notwithstanding, E. O. Wilson, prophet of a middle road, argues that there are two sorts of people, empiricists and transcendentalists, and that the robustness of this bi-polarity may be resolved "by the continuance of biological studies of complex human behaviour."[64]

Theologians from the most conservative to the most liberal have struggled to explain – or even merely to describe – the special nature of humankind in the face of evolutionary science. Humans are obviously different from the apes in many ways – but are the differences merely ones of degree or is there a real qualitative difference? This may well be unanswerable from the scientific point of view but, in the Abrahamic faiths at least, the biological species *Homo sapiens,* descended from a primitive simian stock and related to living apes, is assumed to have been transformed by God at some time in history into creatures which can relate to him.[65] The only alternative to this is to opt for a wholly naturalistic – which tends to imply atheistic – origin for humankind. This is obviously a key issue for anyone with religious beliefs: are we nothing but apes or are we more than apes? If we reject the reductionism of "nothing-but-an-ape", we can conveniently call the transformed ape *Homo divinus,* biologically unchanged but spiritually distinct.[66] We are apes, but we are more than apes and it is useful to mark this difference with a change of name.

Is it possible to say anything about this transformation, other than accepting that at some time in history an ape became a human? Its timing and consequences have been elusive. At one time, humans were believed to be the only creatures able to walk on two legs. When this was shown to be false, there was a phase of identifying humanness with tool-using, then tool-making and storing, then intelligence or reason, then self-awareness, and perhaps most persistently as the only possessors of a 'soul'. Genesis 1 describes the appearance of what I am referring to as *H. divinus* in terms of humankind being made in 'God's image', by a specific act of God;[67]

---

[64] Edward O. Wilson, *Consilience* (New York: Knopf, 1998), 264.

[65] J. Richard Middleton, *The Liberating Image. The imago Dei in Genesis 1* (Grand Rapids: Brazos Press, 2005)

[66] John R.W. Stott, *Understanding the Bible* (London: Scripture Union, 1972); R.J. Berry, "This cursed earth. Is 'the Fall' credible?," *Science & Christian Belief* 11 (1999): 29-49, 165-67; R.J.Berry, "Adam or Adamah?," *Science & Christian Belief* 23 (2011): 23-48

[67] Genesis 1 uses two different Hebrew words to describe God's creative work: yasah (which has the sense of modelling from previous material, as a potter moulds

Genesis 2:7 describes this as a divine in-breathing into an already existing entity. Scriptural exegesis and modern neurobiology join in rejecting the notion that this is the addition of a soul to a body.[68] The modern consensus is that the possession of God's image implies a function not an entity. For H.D. McDonald, "image should be taken as indicating sonship," seeing this as holding together both the ontological and relational aspects of the image.[69] Claus Westermann suggests that "the relationship to God is not something which is added to human existence; humans are created in such a way that their very existence is intended to be their relationship to God."[70] The 'image of God' means that we are sufficiently like God that we can have an intimate relationship with him: we are told that God walks

---

clay) and bãrã which is used to refer to the creation of matter (v.1), the great sea monsters (v.21) and mankind (v.26), and which is always used in the Bible to refer to God's creative activity. It would be dangerous to build too much on this use of words. Both refer to a divine work and there is no clear demarcation in the Bible between God's works in nature and his works in history; he is sovereign in both. Notwithstanding, as Westermann comments, "the meaning is indisputable. Man in his entirety … is to be designated as a creature in God's image. … [It] serves to underline the uniqueness of man's creation. The creation of man is something far different from the creation of the rest of the world. One can almost say that this ruptures the framework of the course of creation in which all the other works of creation are included" (*The Genesis Accounts of Creation*, Philadelphia: Fortress, 1964, 21).

[68] Warren S. Brown, Nancey Murphy, and H. Newton Malony, eds., *Whatever Happened to the Soul?* (Minneapolis: Fortress Press, 1998); Malcolm Jeeves, "Towards a composite portrait of human nature," in *From Cells to Souls – and Beyond*, ed. Malcolm Jeeves (Grand Rapids: Eerdmans, 2004), 233-49; Ralph F. Stearley, "Assessing evidences for the evolution of a human cognitive platform for 'soulish behaviors,'" *Perspectives on Science and Christian Faith* 61 (2009): 152-74. David Fergusson quotes von Rad on the meaning of image: "Just as powerful earthly kings, to indicate their claim to dominion, erect an image of themselves in the provinces of their empire where they do not personally appear, so man is placed on earth in God's image as God's sovereign emblem." Fergusson continues, "The function of the human being is to exercise God's rule over the non-human creation. … The image of God is to be understood not substantively in terms of the possession of an immortal soul, but relationally in terms of the role that human beings play before God and before the rest of creation." David Fergusson, *The Cosmos and the Creator* (London: SPCK, 1998), 14.

[69] H.D. McDonald, *The Christian View of Man* (London: Marshall, Morgan and Scott, 1981), 39-41.

[70] Claus Westermann, *Genesis 1-11*, trans. J.J. Scullion (London: SPCK, 1984), 158.

in the garden with Adam and Eve, and he speaks to them personally – in a different way to the rest of creation (Genesis 1:28).

Can we say anything further about the instantiation of humanness (*i.e.* the acquisition of the *imago Dei*)? (I use the word 'instantiation' to avoid the mechanical implication of words such as 'insertion' or 'implanting'.) Science can contribute on two points. The '*imago*' is as intrinsic a part of humanness as eye colour or memory, but it would be wrong to assume that it is inherited or transmitted like a Mendelian trait. This means that if it was initially conferred on an individual, there is no reason why it should not spread by divine *fiat* to all other members of *Homo sapiens* alive at the time. In the same way, the effects of the disobedience of the first pair could also have spread 'laterally', a suggestion made by Derek Kidner in his Tyndale Commentary on Genesis.[71] This understanding would emphasize the responsibility of individuals to respond to God in obedience, whereas interpreting 'Adam' as corporate humanity reduces personal commitment and the consequences of disobedience.

There is no reason to insist that the 'instantiation' event took place at the same time as the emergence of the biological form we call *H. sapiens*; it was not a genetic change. Nor does it seem to matter whether this was a sudden change or one that took place over time (as an emergent character). God's acts are not always instantaneous (*e.g.* when Jesus healed a blind man, Mark 8:23-25). Secondly, the 'death' that entered the world "the day

---

[71] Derek Kidner, *Genesis* (London: Tyndale, 1967), 29. This suggestion means also that the possibility of an original single pair (i.e. an historic Adam and Eve) is not irrational. It is certainly implied by Paul's repeated insistence on 'one man' in Romans 5:6-19. Leon Morris (*The Epistle to the Romans*, Leicester: IVP, 1988, 228) asserts: "the one man [Adam] is very important and underlies the whole discussion [in Romans, Chapter 5]. Twelve times in verses 12-19 we have the word one; repeatedly Paul refers to the one man Adam (and to one sin of that one man) and opposes to him (and it) the one man Jesus Christ (and his one word of grace). The one man and his sin and the one Saviour and his salvation are critical to the discussion." (See also John R.W. Stott, *The Message of Romans*, Leicester: IVP, 1994, 163). Notwithstanding, Morris warns "Paul's argument in Romans 5 is very condensed and in all translations and comments we must allow for the possibility that Paul's meaning may at some point be other than we think." Intriguingly, James Barr (*Biblical Faith and Natural Theology*, Oxford: Clarendon Press, 1993, 167) also argues for the possibility of a historic Adam. He points out that Paul makes much of the distinction in Gal. 3:16 between 'seed' or offspring (singular) and 'seeds' (plural), and hence he would be unlikely to mean men (plural) in Romans 5 where he repeatedly speaks of 'one man'. The historicity of Adam and Eve is not important in the context of this essay, but it is worth noting that its rejection on the grounds of scientific implausibility is not correct.

that Adam sinned" was, of course, primarily separation from God (*e.g.* Ephesians 2:1-5), a rupture of relationship. Physical death was part of creation from the beginning, not least in the Creator's provision of plants for animals to eat (Genesis 1:29, 30; 2:16).

Up – or Down – from the Ape?

Peter Bowler has well captured a major difficulty about interpreting humanity as the result of unqualified Darwinian evolution:

> In the rush to abandon biblical literalism, vital doctrinal problems have been evaded rather than confronted. The most obvious source of friction was the traditional belief that the human race had fallen from an original state of grace, redemption by Christ being the only source of salvation. If evolutionism was true – even in its progressionist and teleological form – there could be no Fall and hence no need for salvation in the original sense of the term. The Fall had to be reinterpreted as the emergence of a moral awareness among early humans.[72]

In its less nuanced American form, this becomes: "No Adam, no Fall; no Fall, no Atonement; no Atonement, no Saviour."

This is where science-faith dialogue comes against the 'grand narrative' of traditional Christianity. Science can give an account of human evolution without invoking God, but this does not mean that there is no God.[73] This is particularly acute when we seek to understand the "temptation story" in Genesis 3, particularly if we treat it as merely a way of describing the advent of self-awareness: that the eating of the fruit which led to "the opening of the eyes" of our first parents means the development of consciousness and the possibility of moral choice. In these terms, the 'Fall' becomes an upward leap into true humanness rather than a descent into moral turpitude. Julian Huxley used to speak of us having moved from the biological phase of evolution into what he called the 'psychosocial phase'; C.H. Waddington described it as the 'sociogenetic phase'. Teilhard de Chardin's idea of progress towards an 'omega point' expressed a similar notion.

---

[72] Peter J. Bowler, "Evolution and the Eucharist: Bishop E.W. Barnes on science and religion in the 1920s and 1930s," *British Journal for the History of Science*, 31 (1988): 454

[73] For example, John Durant, ed., *Darwinism and Divinity* (Oxford: Basil Blackwell, 1985); Mary Midgley, *Evolution as a Religion* (London: Methuen, 1985); Patricia A. Williams, *Doing without Adam and Eve* (Minneapolis: Fortress Press, 2001).

In contrast, the Bible and the Qur'an describe humankind as innately helpless and disobedient to its maker. The essence of the Genesis 3 story is a break in relationship between the creator and human creature through disobedience, an event which we call 'the Fall' and which resulted in 'death', that is an inability to relate to God. Science can say nothing about God, and hence can do no more than speculate about this. Christians go further than Jews or Muslims and insist that the answer to the Fall is that Christ made possible the restoration of the original relationship by his redeeming death on the cross (Colossians 1:20).[74] This reconciliation is expressed in various ways: Jesus told Nicodemus that he must be "born again" by the Spirit, not by a biological process (John 3:3-18); Paul told the Ephesians, "You were dead [but God] brought us to life with Christ when we were dead" (Ephesians 2:1-5); for John, "We know we have crossed over from death to life" (1 John 3:14); for the author of the Revelation, we have a choice as to whether we "open the door" to the Christ (Revelation 3:20). Is there any way that this theme – which runs throughout the Bible – can be reconciled with the scientific story of evolution from the ape?

One way is to ignore the Fall narrative on the grounds that it does not seem to be possible to interpret it literally and it is hardly touched upon in the rest of the Bible. Whilst both points are true, this assumption neglects the main thrust of scripture – that of a loving and caring God who repeatedly rescues his disobedient people. Significantly too, it does not take into account the implications of the transformation of *Homo sapiens* into *Homo divinus*.

The fullest references to the Fall in the New Testament are in Romans 8:18-22, where the created universe is said to have been "made subject to frustration" and to be "groaning as if in the pangs of childbirth." Charles Cranfield has used a *reductio ad adsurbum* argument to point out that the frustration suffered by the creation can only have meaning if it refers to a failure of "the united praise of the whole creation" when *we* "fail to play our rational part … just as all the other players in a concerto would be frustrated of their purpose if the soloist were to fail to play his part."[75] Commenting on the Genesis story, Derek Kidner uses a similar analogy to link the pre-Fall situation with our present existence:

---

[74] Some of the ideas raised by this context have been explored by Celia Deane-Drummond, *Christ and Evolution* (Minneapolis: Fortress Press, 2009).

[75] Charles Cranfield, "Some observations on Romans 8:19-21," in *Reconciliation and Hope: New Testament Essays on Atonement and Eschatology presented to L.L. Morris on his 60th Birthday*, ed. R. Banks (Grand Rapids: Eerdmans, 1974), 227.

> Leaderless, the choir of creation can only grind on in discord. It seems from Romans 8:19-23 and from what is known of the pre-human world that there was a state of travail from the first which man was empowered to 'subdue' until he relapsed into disorder himself.[76]

Blocher makes essentially the same point:

> If man obeys God, he would be the means of blessing the earth, but in his insatiable greed … and in his short-sighted selfishness, he pollutes the earth and destroys it. He turns a garden into a desert (cf. Rev 11:18). That is the main thrust of the curse of Genesis 3.[77]

For C.F.D. Moule, the Romans passage means that:

> … man is responsible before God for nature. As long as man refuses to play the part assigned him by God, so long the entire world of nature is frustrated and dislocated. It is only when man is truly fitting into his proper position as a son in relation to God his Father that the dislocation of nature will be reduced.[78]

The possession of 'God's image' is the mark of our distinctiveness as humans. But God's image is a divine attribute. It is not an evolved character. Although its expression depends on our bodies, it is not genetic. This reading of scripture is important, because it emphasizes the extraordinary providence (and sovereignty) of God while contrasting radically with ideas that the future of the world is built upon a sophisticated version of deism. The latter supposition implies that creation continues as some sort of intrinsic unfolding. It is theologically seductive and has repeatedly surfaced in the past 100 years or so in philosophical speculation. It takes us back to the *élan vital* of Henri Bergson, and is reflected in the *noogenesis* of Teilhard de Chardin, the panentheism of Hartshorne, Cobb and Peacocke, Lovelock's Gaia, the 'promise of nature' as advocated by John Haught,[79] and 'strong emergence' as proposed by

---

[76] Kidner, *Genesis*, note 64, 73.

[77] Henri Blocher, *In the Beginning: The Opening Chapters of Genesis* (Leicester: IVP, 1984), 184.

[78] C.F.D. Moule, *Man and Nature in the New Testament* (London: Athlone, 1964), 12.

[79] John Haught, J. (2005) "The Boyle Lecture 2003: Darwin, design and the promise of nature," *Science & Christian Belief* 17 (2005): 5-20.

Philip Clayton.[80] It is pertinent to recall Aubrey Moore's welcome, in the early days of post-*Origin* debates, of the banishment of insidious deism because Darwinism "under the disguise of a foe, did the work of a friend." His argument was that early nineteenth century deism had left God "throned in magnificent splendour in a remote corner of the universe" and a machinery of "secondary causes" had taken his place:

> Science had pushed the deist's God farther and farther away [until] at the moment when it seemed as if he would be thrust out altogether, Darwinism appeared and … conferred upon philosophy and religion an inestimable benefit by showing that we must choose between two alternatives. Either God is everywhere present in nature or He is nowhere.[81]

The Orthodox emphasis on creation as gift is helpful here, because of the light it throws on the nature of the giver.[82] The most robust model for God's action in the world is complementarity.[83]

From the present point of view, a rewarding approach seems to be to follow the long-standing tradition that God wrote two books: a book of words (the Bible) and a book of works (creation). The books have the same author but are written in very different languages and therefore have to be understood through entirely different techniques. Notwithstanding, any contradiction between the two must be false. On the title page of the *Origin of Species*, Darwin quoted from Francis Bacon, "Let no man think or maintain that he can search too far or be too well studied in the book of God's words or in the book of God's works; but rather let men endeavour an endless … proficience in both." Presumably it was a discipline he endorsed. Near the end of his life, Darwin wrote to his friend Brodie Innes, the former Vicar of Downe, "I hardly see how religion and science can be kept separate," and to the sceptic John Fordyce, "It seems to me absurd to doubt that a man may be an ardent Theist and an evolutionist."

---

80 Philip Clayton, *Mind and Emergence: from Quantum to Consciousness* (New York: Oxford University Press, 2005).

81 Aubrey Moore, "The Christian doctrine of God," in *Lux Mundi*, ed. C. Gore (London: John Murray, 1889), 99-100.

82 John Zizioulas, J. (2005) "Creation theology: an Orthodox perspective," in *Listening to Creation Groaning*, ed. L. Vischer (Geneva: John Knox International Reformed Center,) 90-104; John Zizioulas, "Priest of creation," in *Environmental Stewardship*, ed. R.J. Berry (London: T&T Clark, 2006), 253-60.

83 Donald M. MacKay, *Behind the Eye* (Oxford Blackwell, 1991); R.J. Berry, *God's Book of Works* (London: T&T Clark, 2003).

Evolving Darwinism

To some, Charles Darwin provided the nails for God's coffin; for others, he described God's work in creation and enabled them to "think God's thoughts after him." Thomas Henry Huxley, Darwin's 'bulldog', notoriously wrote, "Exhausted theologians lie about the cradle of every science as the strangled snakes beside that of Hercules; and history records that whereas science and orthodoxy have been fairly opposed, the latter has been forced to retire from the lists, bleeding and crushed if not annihilated; scotched if not slain."[84] Intriguingly, the same bruiser wrote later in life:

> It is the secret of the best theological teachers to the majority of their opponents that they substantially recognize the realities of things, however strange the forms in which they clothe their conceptions. The doctrines of predestination; of original sin; of the innate depravity of man and the evil fate of the greater part of the human race; of the primacy of Satan in this world, faulty as they are, appear to me to be vastly nearer the truth than the 'liberal' popular illusions that babes are all born good and that the example of a corrupt society is responsible for their failure to remain so; that it is given to everybody to reach the ethical ideal if he will only try.[85]

Huxley's comments were – as always – robust, but he was not contradicting himself in these assertions. The *Origin* forced a re-aligning of faith and reason in ways unexpected by most, not least Darwin himself. It made sense of a vast amount of apparently disparate data on anatomy and physiology, animal and plant distribution, and ecological processes, which together have provided a basis for scientific advances as far-reaching as relativity theory or the Periodic Table. It also sounded the death knell of deism and the opening of new ways to explore the relationship of the Creator to the creation.

Darwinism made it possible to think of a world without God, a world ruled entirely by 'natural processes'. This is the choice which allowed Dawkins to proclaim that "although atheism was logically tenable before Darwin, Darwin made it possible to be an intellectually fulfilled atheist."[86] If decision about the existence or not of God depended on natural theology – the old evidence of 'design' – the case for theism would be seriously weakened. But Christianity is based on the historical coming of the Christ

---

[84] T.H. Huxley, "Review of the Origin of Species," *Westminster Review* 17 (1860): 556.

[85] T.H. Huxley, "An apologetic irenicon," *Fortnightly Review* 52 (1892): 569.

[86] Richard Dawkins, *The Blind Watchmaker* (London: Longman, 1986), 6.

and his resurrection from the dead. The Christian faith does not stand or fall on inferences from the natural world; evolution and Darwinism buttress faith, they neither prove nor banish it. This is the context in which we should site Darwin.

Sadly, debates about evolution and creation tend to divert efforts away from building a robust and refreshed natural theology. This is a tragic legacy of Darwinism. The challenge now is to move on from the kick-start of 1859 and develop a strong doctrine of creation. Can we rise above the utilitarianism of reacting to issues of climate change, population pressure, resource depletion, and the like? Lynn White stirred many with his criticism of unthinking human reliance on technology in his essay on the 'Historical roots of our ecologic crisis.'[87] Recent decades have seen a considerable rise in Christian concern for and action over the environment, but creation-care still remains a minority passion for enthusiasts, rather than an integral part of Christian belief. Is it too simplistic to see this failure of understanding as at least in part a reaction against the evolution-creation struggles – a reaction against the naiveties of both Darwinian atheists on the one hand and the "creationists" on the other?

[87] Lynn White, "The Historical Roots of Our Ecologic Crisis," *Science* 155 (1967): 1203-07. White has been criticized by both historians and theologians: see, for example, Joseph K. Sheldon, *Rediscovery of Creation: a Bibliographic Study of the Church's Response to the Environmental Crisis* (Metuchen, NJ: Scarecrow Press, 1992); R.J. Berry, ed., *The Care of Creation* (Leicester: IVP, 2000); but his main thesis of human arrogance and selfishness stands.

Chapter Three

Response: *Homo Divinus* – Myth or Reality?

Celia Deane-Drummond

R.J. Berry's paper takes us to the heart of what might be termed Charles Darwin's lifeworld.[1] As well as re-creating some fascinating and illuminating historical details relating to Darwin's social, religious and cultural sphere, I particularly appreciated Berry's insight into how Darwin thought *as a biologist and naturalist.* Berry traces how Darwin's views of the adequacy of scientific and theological explanation gradually changed in response to particular observations, as those of any good biologist would. He takes us back to the historical and cultural details of Darwin's day, when the belief in God as a remote Deity was popular. In this perspective, Darwin did theology a favour, for it was no longer possible to consider God as a distant God who simply intervened at intervals in creation; rather, if Darwin was right, God had to be re-conceived as intimately and immanently involved in the creative process. This is precisely what Henry Drummond and others like him attempted to do in the aftermath of Darwin's *Origin of Species.* Darwin's biography shows him up as first and foremost a naturalist, which, in his day meant one who was capable of intense observation of the natural world and meticulous attention to detailed recording. His modesty is also striking in the excerpts Berry highlights for us, for the judgement of his teachers that he was of "lesser intelligence" in his formative years seems to have had a lasting impact. But Darwin was also prepared to be bolder in his letters to his

[1] This is a term used by Jürgen Habermas to describe a perspective from the inside, rather than the outside. See, for example, J. Habermas, *The Theory of Communicative Action: Vol. 2 Lifeworld and System, A Critique of Functionalist Reason*, trans. Thomas McCarthy (Cambridge: Polity, 1987).

intimates, and these reveal a growing confidence against rival theories, such as that proposed by Lamarck.

Darwin's odyssey

What is fascinating about Berry's account is not simply the struggle for ascendency among rival biological theories at the time, but the psychology of Darwin's odyssey. Added to this, we have insights into the scientific currents that existed alongside Darwin's thesis that reinforced the gradual erosion of Paley's designed world – what might be called 'deep space' and 'deep time' and 'multiple origins' were also all in the air. Berry shows us precisely where Darwin's genius lay – it was in linking known facts about population growth and the struggle for existence with the idea of heritable variation. But what might seem a relatively small logical step had immense repercussions. All this is situated in the context of Darwin's personal life – the deep faith of his wife impacting on the timing of the publication of the *Origin*, compounded by a concern not to repeat the errors of Robert Chambers' *Vestiges,* alongside personal modesty about the depth of his own research. But another rival biologist, Alfred Russel Wallace, jolted Darwin out of a misplaced sense of complacency – perhaps omitted by Berry in this account as this aspect is so well rehearsed. Added to this he discovered flaws in the inductive method that forced him to consider a methodological alternative less well-trodden at the time, hypothetico-deduction. All these details recreate the drama of Darwin's odyssey and the seeming fortuitous events that influenced the way the *Origin* was conceived and the manner in which it was eventually released. Berry also treats the theological reaction to Darwin with a degree of care uncommon among his interpreters, especially among conservative evangelicals. Berry lets such historical facts speak for themselves – showing up the possibility of a liaison between conservative views and evolutionary theory by clear historical example.

The reception history of Darwinism

Yet Berry also takes us in this essay into what might be called the *reception history* of Darwin's hypothesis among biologists. This is important, since it is too easy to slip into the comforting conclusion that the only scholars who objected to his theory were theologians or those of religious persuasion. Instead, by carving out those objections and watching the gradual rehabilitation of Darwin's views, albeit in a modified form, Berry gives us a window into the historical process of science. In doing so he takes us right up to contemporary debates in paleobiology over the degree of convergence in evolutionary processes, and therefore the extent of restriction on the processes of natural selection.

For Berry the biggest theological challenge is not so much creation, but theological *anthropology*: how to envisage in theological terms the place of human beings. I agree with his interpretation of Darwin in as much as, in contemporary discussion, this is still an area of heated debate. Berry's welcome summary of hominid evolution is a candid reminder to would-be doubters that the evidence for human evolution is virtually uncontested. Further, in spite of the fact that humans separated from chimpanzees some five million years ago, over 95% of the genome is common to both species. This is remarkable genetic conservation, though the differences in gene regulation are striking, as is the fact that chimpanzees seem to be more adapted to their environment than humans; so just a tiny genetic change could have led to the possibility of highly developed language skills in humans.

When it comes to a discussion of socio-biology Berry seems to hesitate to position himself; naming E.O. Wilson as a 'prophet of a middle road' (p. 30). Yet while Berry is reluctant to declare that human beings are described solely by different shades of biology and evolution, he leaves the criticism of socio-biology undeveloped, implying acceptance. Of course, we reach a dilemma here that is not properly aired, namely, that while we are clearly creatures of evolutionary origin, our capacities and cultural communities are not so readily explained by biological accounts. The question becomes *how precisely* do biology and cultural factors shade into one another in the assessment of the human condition? E.O. Wilson, who claimed, at least in his earlier work, to keep culture 'on the leash' of our genes, does not, it seems to me, offer a prophetic middle way in the manner that Berry implies.

Homo Divinus?

And then we have what strikes me as the even more astonishing claim, following J.R. Middleton, that according to the Abrahamic faith God "transformed humans" at some point in their evolutionary history as *Homo sapiens*. For Berry, "The only alternative to this is to opt for a wholly naturalistic – which tends to imply atheistic – origin for humankind" (p. 30). But how is this consistent with Berry's claim earlier in his essay that "There is no scientific support for the notion that we have been propelled towards a predetermined end by a Divine Watchmaker" (p. 27)? Transformation could, of course, not involve any pre-set end, but it implies that there is one, especially when set against the belief in an alternative, "wholly naturalistic" origin.

Why do I find this astonishing? I find it astonishing because it seems to presuppose that the account in the book of Genesis has to be taken at face value, that God was specifically involved in the creative act of making humans in a way that is not true of other species. But this sets up a divide that splits humans apart from their evolutionary history, and makes God into a God who acts by direct intervention, rather than through evolutionary events. If Berry finds a stumbling block here, then what about the whole history of life, and evolution in particular? Why is it so problematic to believe that God works for the human species in evolution? This sets up a barrier between humans and other animals which smacks of the anthropocentrism that has unhelpfully dominated the history of Christian thought.[2] In other words, the idea of God intervening at some point in the history of *Homo sapiens* seems to go against the sufficiency of God's presence in the world through divine Providence, and to speak of an intervention at some gap-point in our scientific knowledge. While he resists the crude notion that this takes the form of the soul being mechanically added to the body, he still seems to assume it is a special work of God at a particular period of history, taking *Homo* from *sapiens* to *divinus.*[3] Berry justifies this development by claiming that "We are apes, but we are more than apes and it is useful to mark this difference with a change of name" (p. 30).

One difficulty that Berry faces is that on purely historical grounds it seems that *Homo sapiens* existed as a biological species long before cultural evolution and any religious belief was in evidence. Perhaps this is the rationale for a change of name, but it implies that *Homo sapiens* prior to this transformation was not really what God intended; or rather, that only those who were able to respond in a manner appropriate to this stage of their cultural history fulfilled the divine intention. But can we really put ourselves in the place of God and decide who or what humans among *Homo sapiens* are worthy of the name human? Berry seems to be creating two subspecies of humans, *H. sapiens* and *H. divinus*, with the former inferior to the latter. But dividing humans up on the basis of purported historical religious qualities has serious social and moral consequences, as

---

[2] For discussion on this and other related matters see Celia Deane-Drummond and David Clough, *Creaturely Theology: On God, Humans and Other Animals* (London: SCM Press, 2009).

[3] The idea of *Homo divinus* is also used by John Stott, who seems to be the first person to use the term. John R.W. Stott, *Understanding the Bible* (London: Scripture Union, 1972), 63.

it opens the door for possible discrimination against other human beings transformed from *sapiens* to *divinus* to a greater or lesser extent. Berry is not intending this development, but I am suggesting that it permits this possibility, and therefore has ethical implications, much as Darwin's categorisation of other humans as 'primitive' provided the logical ground for the persistence of discriminatory practices towards other peoples, including the institution of slavery.

Of course, Berry is on safe ground in that understanding fully what happened in the course of *H. sapiens'* evolutionary history is unlikely to be achieved; given our time and distance from when this took place, all we see are hints at religious belief. But then, when might we recognise this as divine intervention? Is it when particular belief in divine agency takes place? Or is it when there is belief in polytheism? Or monotheism? Are we saying that God introduces humanity deliberately to superstition? Darwin, in his *Descent of Man*, was well aware of the differences between these forms of belief; and, while he argued that belief in some sort of religious agency may have evolved, he distinguished this from belief in an Omnipotent God. In his chapter on mental powers he speaks about belief in religion in the following way:

> There is no evidence to suggest that man was aboriginally endowed with the ennobling belief in the existence of an Omnipotent God. On the contrary, there is ample evidence, derived not from hasty travellers, but from men who have long resided with savages, that numerous races have existed, and still exist, who have no idea of one or more gods, and who have no words in their language to express such an idea. The question is of course wholly distinct from that higher one, whether there exists a Creator and Ruler of the universe, and this has been answered in the affirmative by some of the highest intellects that have ever existed. If, however, we include under the term 'religion' the belief in unseen or spiritual agencies, the case is wholly different; for this belief seems to be universal with the less civilized races (142-3). … The belief in spiritual agencies would pass into the belief in the existence of one or more gods (145). … The same high mental faculties which first led man to believe in unseen spiritual agencies, then in fetishism, polytheism and ultimately monotheism, would infallibly lead him, as long as his reasoning powers remained poorly developed, to various strange superstitions and customs (146). [4]

The question that confronts us here is how far is belief in divine transformation in the history of *Homo* credible in the light of evolutionary history? Or should we seek another explanatory route, using anthropological

[4] Charles Darwin, *The Descent of Man and Selection in Relation to Sex* [1871] (London: Penguin, 2004).

and philosophical analysis? I agree with Berry that human beings are distinct biologically, and, as self-aware religious beings they are distinct theologically. But attempting to map the Genesis account of the creation of humanity directly onto evolutionary history does not work as far as I am concerned, any more than forcing the days of the first Genesis creation story into the Big Bang and subsequent evolutionary accounts.

Evolution and the Fall

Part of the struggle that Berry has here is reconciling a particularly conservative view of the biblical record with the history of humankind. Hence, "In the same way, the effects of the disobedience of the first pair could also have spread 'laterally', a suggestion made by Derek Kidner in his Tyndale Commentary on Genesis" (p. 32). I suspect that the conservative evangelical insistence on the historicity of the first human pair (though clearly not the first *Homo sapiens*) is misplaced, for the Garden of Eden is an evolutionary impossibility, so it would imply a historical pair in the midst of a phantom world. I, like Berry, believe that the Genesis account of the first humans and their Fall may still be retrievable; not, however, for me by a direct mapping onto evolutionary history, but by a re-interpretation of its meaning. For example, if the significance of the Fall is rooted in mythology, or pre-history, then in the light of evolutionary history it makes more sense to broaden the Fall so that it can be interpreted in a more inclusive way as having premonition in the history of social primates, or even earlier hominids, rather than having occurred at a fixed historical point. I do not, therefore, feel squeamish about doubting the literal origin of the figures of Adam and Eve in the book of Genesis; rather, this tale is one that speaks to its readers of the presence of God from the earliest moment of human history.

For Berry "interpreting 'Adam' as corporate humanity reduces personal commitment and the consequences of disobedience" (p. 32). But why should this necessarily be the case? If the story of the Fall has to be taken literally, and Adam equated with Neolithic man, then so does the idea that humans are made out of dust, which clearly does not make much sense in the context of evolutionary knowledge. Rather, knowing God's presence with humans from the earliest moments of human existence should encourage commitment, for God is "with us" in our pre-history, and commitment – at least in Christian terms – comes from knowing the salvation from sin possible in Jesus Christ, rather than the facticity of Adam's creation and Fall.

The interpretation of the early Genesis texts that I am seeking to develop is different from other, more liberal attempts to render the Fall in largely positive terms as a fall 'upwards' into self awareness, or to dismiss the theological significance of the Fall altogether; these rejections I share with Berry. Rather, it is a sinking that reaches its climax in humankind, now become aware of God and recognising a broken relationship. The choice is not, as Berry implies, between corporate humanity and a literal Adam; rather, humans need to be understood as a social species from the beginning. While interpretations of the Genesis account according to 'corporate humanity' may be detached from a belief that the fall of humanity represented something historical, it is not strictly necessary to do this. The idea of an evolutionary account of human fallenness that I am arguing for here is similar in some respects to the position of Jesuit scholar Raymund Schwager,[5] except that I perceive the possibilities for cooperation or its opposite as existing in other social species, and therefore prior to hominid evolution.[6] Schwager argues for a fall of humanity at the dawn of human consciousness. This allows a preservation of historicity in the Genesis account, but resists any literalism in the story of Adam and Eve, who are what might be termed figures *representative* of this pre-history. I suggest that Berry's approach to the Fall as adhering to a literal Adam and Eve is unnecessary even from an evangelical perspective. Careful biblical scholarship supports the idea of both a generalised account of the Fall, such that it spread into the human condition as such, and an admission of some basis in history, though not necessarily a literal one.[7]

Berry claims, further, that "God's image is a divine attribute. It is not an evolved character. Although its expression depends on our bodies, it is not genetic" (p. 37). I agree with this statement in as much as image-bearing is of far more than bodily origin, but there is no evidence to suggest that image-bearing is somehow 'implanted' into humans by God from the outside, or that it is split off from our evolved bodyliness. Rather, the most convincing accounts of image-bearing relate to the specific *task* of human beings in exercising their responsibility for the earth, rather than a specific *attribute*, divine or otherwise. In the history of Christian thought, image-bearing has been thought of as an ontological characteristic, often

---

[5] Raymund Schwager, *Banished from Eden: Original Sin and Evolutionary Theory in the Drama of Salvation* (Leominster: Gracewing, 2006).

[6] For further discussion see Celia Deane-Drummond, *Christ and Evolution: Wonder and Wisdom* (Minneapolis: Fortress; London: SCM Press, 2009).

[7] For a helpful discussion from an evangelical perspective see John J. Bimson, "Doctrines of the Fall and Sin After Darwin", in *Theology After Darwin*, ed. Michael S. Northcott and R.J. Berry (Milton Keynes: Paternoster, 2009), 106-122.

named as 'reason' or 'intelligence'. Berry seems to connect image-bearing with divine inbreathing and even a God-given capacity for religious belief, but this is not implied in the text.[8] In other words, Berry's exegesis of what image-bearing might entail is extremely misleading, and gives the impression that image-bearing is the work of a religious soul, somehow added to bodily humanity, whose reliance on bodily capacity relates to 'expression' rather than being grounded in bodily being as such. This separation of body and spirit implied by Berry's account fails to portray an acceptable meaning of image-bearing. Instead, image-bearing means that something of the character of God is echoed in the way humans act that is not present in other animals. This strikes me as close to the Hebrew sense of what it means to be human. In other words, the functional aspect of image-bearing is more convincing, rather than relating this to what amounts to a dubious form of religious ontology.

Conclusion

In conclusion, Berry's dive into the historical aspects of Charles Darwin's life and its immediate aftermath presents us with extremely valuable insights into Darwin as a biologist in the particular religious and cultural world of his age. The lessons from history are relevant today as they show that it is possible both to be a biologist accepting Darwin's theory of natural selection *and* hold to a conservative view of religious belief. There are, therefore, no grounds to resort to either rejection of Darwin's views in the name of the Bible, or adherence to creationist accounts. However, this belief does not seem to be followed through consistently in Berry's treatment of anthropology. His argument for *Homo divinus* is unconvincing, not just from a biological perspective but also exegetically and theologically, and implies an unhelpful impression of a God who intervenes directly in human history in a manner that Darwin's account helped to reverse. While Berry certainly rejects deism in all its forms, the logical extension of *H. divinus* implies division of the human species into those who win God's favour and those who do not. This has serious moral implications, given the way that Darwin's theory has already been abused in human history.[9]

---

[8] Bimson also offers critical comments on Berry's exegesis linking divine breath with image-bearing. See Bimson, "Doctrines of the Fall", 114-115.

[9] For a discussion of the interpretation of Darwin that led to myths about favoured races, see Amy Laura Hall, "Charles Kingsley's Christian Darwinism", in Northcott and Berry, *Theology After Darwin*, 41-56.

CHAPTER FOUR

CHRISTIAN DARWINIANS

JOHN HEDLEY BROOKE

The conference at which the argument of this essay was first rehearsed had as its subtitle: "From natural theology to a theology of nature." This formula nicely captures an important aspect of the Darwinian impact on Christian theology. Whereas inductive inferences to a Creator from what was known of the natural world had once been commonplace, in a Darwinian universe religious interpretations of nature were more transparently dependent on prior theological assumptions. In 1868, Darwin published his study of *Variation in Animals and Plants under Domestication*. The book contained a provocative analogy that was directed against the claim that the variations on which natural selection worked had been providentially led in propitious directions. A man could build a house from the stones of many shapes and sizes that he might find at the bottom of a precipice. But surely no one would say that the stones had come to be there in order that such a house could be built? Similarly, Darwin claimed, the variations occurring in plants and animals were so randomly distributed that it was impossible to say they had come into being with a prospective use in mind. To claim that they had, in Darwin's view, would render his mechanism of natural selection superfluous.[1] He had in mind his old friend, the Harvard botanist Asa Gray. Nobody in North America had done more than Gray to promote the new theory, for which Darwin was doubly grateful because, unlike several early commentators, Gray really understood what Darwin had meant by natural selection. However, as a Presbyterian Christian, Gray had also wished to

---

[1] Charles Darwin, *The Variation of Animals and Plants under Domestication* (London: Murray, 1868); Frederick Burkhardt, ed., *The Correspondence of Charles Darwin, vol. 16 part 1* (Cambridge: Cambridge University Press, 2008), xxvii; James R. Moore, *The Post-Darwinian Controversies* (Cambridge: Cambridge University Press, 1979), 275-6.

show that natural selection was not inconsistent with natural theology.[2] It is Gray's response to Darwin's analogy that is so revealing, marking perhaps a pivotal moment in the intellectual history of the western world. In a letter of May 25, 1868, Gray conceded the force of Darwin's sceptical analogy: "I found your stone-house argument unanswerable in substance (for the notion of design must after all rest mostly on faith, and on accumulation of adaptations)."[3] It is hard to deny that one of the consequences of Darwin's theory was a wresting of the initiative from a natural theology that had placed a high value on the rational demonstration of God's existence and attributes. Design was now to be perceived primarily by the eye of faith. Gray could still marvel at the results of a process in which the gradual accumulation of minor variations had eventuated in the production of what Darwin himself called the higher animals. But that was a long way from proving providence from nature.

The waning of natural theology

Too much weight, however, can be placed on the impact of Darwin's theory in effecting such a shift. Sophisticated advocates of natural theology, such as the Cambridge polymath William Whewell, had earlier insisted that the conviction of design was closer to the premise of their arguments than a demonstrable conclusion from the evidence. Arguments for design, Whewell implied, would only be persuasive to those who had already cultivated the right frame of mind.[4] The novelist and Christian socialist Charles Kingsley, who was to be one of Darwin's earliest converts, was already confessing to F.D. Maurice in July 1856 that he had long ago found out how little he could discover about God's absolute love, or absolute righteousness, from a universe in which "everything is eternally *eating* everything else." He added that "the study of nature can teach no *moral theology*. It may unteach it, if the roots of moral theology be not already healthy and deep in the mind."[5] A high profile for natural theology faced serious censure from John Henry Newman – not least because he considered it had nothing to teach about the fundamental

---

[2] Asa Gray, *Darwiniana*, ed., A. Hunter Dupree (Cambridge: Harvard University Press, 1963), 72-145.

[3] Burkhardt, *Darwin Correspondence,* vol.16 part 1, 536-7.

[4] John Hedley Brooke, "Indications of a Creator: Whewell as Apologist and Priest," in *William Whewell: A Composite Portrait*, ed. Menachem Fisch and Simon Schaffer (Oxford: Oxford University Press, 1991), 149-73, 154.

[5] Charles Kingsley to F. D. Maurice, July 1856, in F. E. Kingsley, ed., *Charles Kingsley: His Letters and Memories of his Life* (London: Kegan Paul, 1883), 181.

doctrines of Christianity. To attempt a physico-theology was also misguided for another reason. In an Epiphany sermon of 1839, Newman observed:

> … the system of physical causes is so much more tangible and satisfying than that of final, that unless there be a pre-existent and independent interest in the inquirer's mind, leading him to dwell on the phenomena which betoken an intelligent Creator, he will certainly follow out those which terminate in the hypothesis of a settled order of nature and self-sustained laws. It is indeed a great question whether atheism is not as philosophically consistent with the phenomena of the physical world, taken by themselves, as the doctrine of a creative and governing power.[6]

Evangelical commentators could be as critical as Newman. Exponents of natural theology tended to stress the perfections of nature in their celebration of the work of a divine craftsman – a practice at variance with pessimistic readings of the fallen world described in *Genesis*. And as the Scottish evangelical Thomas Chalmers argued in the concluding chapter of his *Bridgewater Treatise*, the most a natural theology could accomplish when addressing the relations between God and humankind was to raise *questions*. For the answers one had to turn to Revelation.[7]

Fifteen years before Darwin published *On the Origin of Species*, the natural theology that Robert Chambers wove into his anonymously published *Vestiges of the Natural History of Creation* (1844) elicited comments to the effect that references to a divine lawgiver had simply become a convenient way of sugaring the pill of scientific naturalism. As James Secord has brilliantly shown, Chambers made the subject of human evolution a matter of wide public debate during the years when Darwin was still perfecting his theory.[8] Denounced by churchmen for its materialism, determinism and immorality, *Vestiges* seemed to point not merely to the limitations of a natural theology but to its vanity. In his review for the Anglican newspaper *The Guardian*, the historian of the Oxford Movement R.W. Church declared that "the *Vestiges* warns us, if proof were required, of the vanity of those boasts which great men used to

---

[6] Cited by Jack Morrell and Arnold Thackray, *Gentlemen of Science* (Oxford: Oxford University Press, 1981), 240-1.

[7] Thomas Chalmers, *On the Power, Wisdom, and Goodness of God as Manifested in the Adaptation of External Nature to the Moral and Intellectual Constitution of Man*, 2 vols. (London: 1833), vol. 2, 282-93.

[8] James A. Secord, *Victorian Sensation: The Extraordinary Publication, Reception, and Secret Authorship of Vestiges of the Natural History of Creation* (Chicago: University of Chicago Press, 2000).

make, that science naturally led on to religion."[9] If, as Jonathan Topham has argued, natural theology had prospered as a vehicle for promoting politically safe science, its eventual waning was a loss to science popularisers and not merely to religious apologists.[10]

These several qualifications are, I think, an embarrassment to Richard Dawkins' assumption that because Darwin undermined William Paley's famous argument for design, he delivered a lethal blow to the foundations of Christianity. It would be going too far, however, to pretend that Darwin's theory was inconsequential for Christian thinkers. Precisely because arguments for design had reinforced perceptions that the sciences were supportive of religion, Darwin's science could be shocking. The dismissive response of Samuel Wilberforce, Bishop of Oxford, has come to symbolise the shock – and the depth of Christian resistance:

> Man's derived supremacy over the earth; man's power of absolute speech; man's gift of reason; man's free will and responsibility; man's fall and … redemption; the incarnation of the Eternal Son; the indwelling of the Eternal Spirit, – all are equally and utterly irreconcilable with the degrading notion of the brute origin of him who was created in the image of God and redeemed by the Eternal Son.[11]

In his published review of Darwin's *Origin*, and possibly in his public dispute with Thomas Henry Huxley, Wilberforce appealed to the scientific authority of Charles Lyell and Richard Owen. This was an invocation of two eminent scientists who had developed distinctive arguments from nature to nature's God and who had considered speculative mechanisms for evolutionary change too premature to warrant acceptance. Lyell had seen evidence of a "Presiding Mind" in the remarkable fact that wherever there was an environment that could sustain life of a particular kind, there that life was to be found. Wilberforce could claim Lyell as an ally because in the second volume of his *Principles of Geology* (1830-33), Lyell had attacked Lamarck's theory of species transformation. In a letter written to

---

9 John Hedley Brooke, "The Natural Theology of the Geologists: Some Theological Strata," in *Images of the Earth: Essays in the History of the Environmental Sciences*, 2nd edition, ed. L. J. Jordanova and Roy Porter (Oxford: Alden Press for the British Society for the History of Science, 1997), 53-74, 68.

10 Jonathan Topham, "Science, Religion, and the History of the Book," in *Science and Religion: New Historical Perspectives*, ed. Thomas Dixon, Geoffrey Cantor and Stephen Pumfrey (Cambridge: Cambridge University Press, 2010), 221-43, especially 223-4.

11 Samuel Wilberforce, "Darwin's Origin of Species," in *Essays Contributed to the Quarterly Review*, 2 vols. (London: Murray, 1874), vol. 1, 52-103, 94.

John Herschel in 1836, Lyell claimed that his idea of a perpetual succession in the extinction and renewal of species "all in accommodation to the changes which must continue in the inanimate and habitable earth" was the "grandest" he had ever conceived "as regards the attributes of the Presiding Mind."[12] Owen had turned to both anatomy and palaeontology for his defence of theism, tracing a common structure in the vertebrates back to an archetypal idea in the mind of God.[13] For Owen the history of life on Earth (and perhaps elsewhere in the universe) was consistent with the unfolding of a divine plan in which there were multiple, and conceivably inexhaustible, instantiations of the archetype. There is no concealing the fact that Darwin's science threatened such constructs and the rhetorical use of them by conservative Christians like Wilberforce.[14]

Diversity and divisiveness

In reputable accounts of Christian reactions to evolutionary biology the diversity of response receives due emphasis – rightly so, given the need to correct popular caricatures that assume nothing but opposition to Darwin from religious interests. During the Darwin bicentenary year it was encouraging to see the popular science writer Steve Jones acknowledging that, far from being uniformly opposed to science, and to Darwin's science in particular, the Anglican Church "soon accommodated Darwin's ideas, which, as most clerics realised, have no relevance to religion."[15] But may this not also be an oversimplification? Stressing the importance of local contexts in determining how religious reactions were shaped, David Livingstone has shown how, even within one and the same theological tradition, a marked heterogeneity of response can be detected. He has a

---

[12] Cited by Edward Bailey, *Charles Lyell* (Edinburgh: Nelson, 1962), 93.

[13] Richard Owen, *On the Nature of Limbs* (London: Voorst, 1849); Nicolaas Rupke, *Richard Owen: Victorian Naturalist* (New Haven: Yale University Press, 1994), 196-7.

[14] For a more comprehensive treatment of the stresses and strains to which natural theology was subject in the nineteenth century, see John Hedley Brooke, *Science and Religion: Some Historical Perspectives* (Cambridge: Cambridge University Press, 1991), chs. 6-8; John Brooke and Geoffrey Cantor, *Reconstructing Nature: The Engagement of Science & Religion* (Edinburgh: T & Clark, 1998; New York: Oxford University Press, 2000), chs. 5-7; Bernard Lightman, *Evolutionary Naturalism in Victorian Britain: The 'Darwinians' and their Critics*, Variorum Collected Studies Series (Farnham: Ashgate, 2009).

[15] Steve Jones, *Darwin's Island: The Galapagos in the Garden of England* (London: Little Brown, 2009), 2.

different story to tell for Presbyterians in Princeton, Belfast, and Edinburgh.[16] Taking diversity seriously means that, in the variety of receptive (and not only oppositional) responses, it is possible to identify issues that were deeply divisive within Christendom. They may not have been destructive but they were damaging. In this essay I want to illustrate both the diversification and the divisiveness of Christian theologies of nature that explicitly addressed Darwinian themes. This is necessary because it would be possible to celebrate the examples of consonance without proper attention to the dissonances between them. So my argument is not that it has been impossible to accommodate Darwinism within Christian theology. It is rather that in their disagreements with each other, Christian Darwinians have revealed just how dangerously divisive Darwin's science could be. In developing this argument, I am following the lead of Bernard Lightman who has suggested that much of the conflict between science and religion in Victorian Britain has to be understood rather in terms of conflict between competing harmonisation strategies.[17]

The divisiveness of Darwin's science is immediately visible in the reactions of two Anglican clergymen, the very two who had been his scientific mentors during his undergraduate years in Cambridge. These were the botanist John Henslow and the geologist Adam Sedgwick. Darwin was hurt when his old friend Sedgwick reproached him for having eliminated final causes from his philosophy of nature. According to Sedgwick, Darwin had destroyed the bond that held together physical and moral discourse.[18] But when Sedgwick spoke out, Henslow leapt to Darwin's defence:

> I got up and stuck up for Darwin as well as I could, refusing to allow that he was guided by any but truthful motives, and declaring that he himself believed he was exalting and not debasing our views of a Creator, in attributing to him a power of imposing laws on the organic world by which

---

[16] David Livingstone, "Replacing Darwinism and Christianity," in *When Science and Christianity Meet*, ed. David C. Lindberg and Ronald L. Numbers (Chicago: University of Chicago Press, 2003), 183-202; and David Livingstone, *Putting Science in its Place* (Chicago: University of Chicago Press, 2003).

[17] Bernard Lightman, "Victorian Sciences and Religions: Discordant Harmonies," *Osiris* 16 (2001): 343-66. See also John Hedley Brooke, "Genesis and the Scientists: Dissonance among the Harmonizers," in *Reading Genesis after Darwin*, ed. Stephen C. Barton and David Wilkinson (Oxford: Oxford University Press, 2009), 93-109.

[18] Adam Sedgwick to Charles Darwin, 24 November 1859, in Burkhardt, *Darwin Correspondence*, vol. 7 (1991), 396-8.

> to do his work, as effectively as his laws imposed upon the inorganic had done in the mineral kingdom.[19]

Just as Henslow disagreed with Sedgwick, so Frederick Temple, a future Archbishop of Canterbury, disagreed with Wilberforce. Preaching in Oxford at the conclusion of the British Association meeting at which Wilberforce and Huxley had their altercation, Temple was far more receptive to the extension of natural laws in the biological domain. Indeed Temple saw theological advantages in bringing the natural order under the reign of law because this rendered it more plausible to believe there is also a moral order to which humans are bound.[20] Temple continued to believe that a unified process of evolution spoke more eloquently of the unity of God than a world in which all living things had been separately created. There were limits to Temple's naturalism when it came to the essence of humankind; but, as early 1860, he was already warning about the dangers of a god-of-the-gaps. Temple and Wilberforce were divided by more than Darwin's science because Temple was a leading contributor to *Essays and Reviews* (1860), which, with its commendation of the historical methods of biblical criticism, drew a vituperative review from the bishop[21] – a reminder that debates about the cultural meaning of science have to be placed in a wider context.

God's immanence and the integrity of nature

Some of the tensions generated by Darwin's science are visible in Darwin himself who admitted that his judgment on matters of religion often fluctuated.[22] He wondered whether it was possible to say that biodiversity was the result of 'designed laws' with the 'details left to chance'. But no sooner do we find him attracted to that formula than he says he is not at all satisfied by it.[23] Underlying such reflections was a deeply divisive issue – whether nature could be assigned an autonomy and

---

[19] John Henslow to Joseph Hooker, 10 May 1860, in Burkhardt, *Darwin Correspondence*, vol. 8 (1993), 200-1.

[20] Frederick Temple, *The Present Relations of Science to Religion* (Oxford: Parker, 1860).

[21] Samuel Wilberforce, *Essays Contributed to the Quarterly Review*, 2 vols. (London: Murray, 1874), vol.1, 114-21.

[22] Charles Darwin to John Fordyce, 7 May 1879, in *The Life and Letters of Charles Darwin*, 3 vols., ed. Francis Darwin (London Murray, 1887), vol. 1, 304.

[23] Charles Darwin to Asa Gray, 22 May 1860, in Burkhardt, *Darwin Correspondence*, vol. 8, 223-4.

integrity such that divine intervention was unnecessary (even demeaning to the deity), or whether Christian theology had such an investment in the miraculous that to outlaw the possibility of divine incursion in the world would be incongruous. This had long been an issue within the Christian tradition, achieving prominence in the Leibniz-Clarke correspondence of the early eighteenth century, and still structuring debates today in the context of 'intelligent design'. Put simply the question is whether the Creator deliberately made a physical world open to divine infusion; or whether the world of nature has what has been called a functional integrity, requiring no such "intervention" to re-stabilise the solar system (in the context of Newton's science) or to generate new species (in the context of Darwin's). There is no doubt about which side Darwin took. To suggest that God could not effect His creative purposes through natural causes he considered almost sacrilegious.[24]

A concomitant issue concerned the balance between divine transcendence and divine immanence. This was another problem that Darwin's theory intensified. It was not difficult to think of a transcendent Creator who had set all the parameters in place such that human beings would eventually emerge. Even Huxley was perfectly willing to countenance what he called a "wider teleology" in this sense.[25] By contrast, there were Christian theologians drawn to Darwin's theory because they saw in a creative evolutionary process evidence of divine participation within the world. For the Oxford theologian Aubrey Moore, Darwin had done Christianity a service by promoting a philosophy of nature that sat comfortably with an Incarnational theology. In a passage from *Lux Mundi* that has appealed to Christian Darwinians seeking a sacramental theology of nature, Moore declared:

> The one absolutely impossible conception of God, in the present day, is that which represents him as an occasional visitor. Science has pushed the deist's God further and further away, and at the moment when it seemed as if He would be thrust out all together, Darwinism appeared, and, under the

---

[24] Dov Ospovat, "God and Natural Selection: the Darwinian Idea of Design," *Journal of the History of Biology* 13 (1980):169-94, especially 183; John Hedley Brooke, "The Relations between Darwin's Science and his Religion," in *Darwinism and Divinity*, ed. John R. Durant (Oxford: Blackwell, 1985), 40-75, especially 46-7; and Brooke, "'Laws impressed on matter by the Creator'?: The Origin and the Question of Religion," in *The Cambridge Companion to the "Origin of Species"*, ed. Michael Ruse and Robert J. Richards (Cambridge: Cambridge University Press, 2009), 256-74.

[25] Thomas Huxley, "On the Reception of the Origin of Species," in *Life and Letters*, ed. Francis Darwin, vol. 2, 179-204, 201.

> disguise of a foe, did the work of a friend ... Either God is everywhere present in nature, or He is nowhere.[26]

There is, however a question whether a sacramental theology of nature has to be such an all-or-nothing affair. When Darwin himself reflected on how his theory might provide a new way of looking at the theodicy problem, he was more discriminating. There were features of nature, in his view, that should not be ascribed to God's immediate presence in all things. There were gruesome contrivances in nature, such as the propensity of the ichneumon wasp to lay its eggs in the body of a caterpillar. Nor could Darwin warm to what he described as a long succession of vile molluscous animals.[27] Their emergence had been possible in a world in which the emergence of human beings had been possible; but it had been one of Darwin's goals to mitigate divine responsibility for suffering and for the more macabre features of creation. So there was, I suggest, a problem with Moore's all-or-nothing claim. Nevertheless, it is certainly possible to argue that, at the close of the nineteenth century, two Christian Darwinians in Oxford, the entomologists Edward Poulton and Frederick Dixey, were sufficiently comfortable with Moore's theology to be uninhibited in their advocacy of natural selection. This was at a time when, as the primary mechanism of evolution, natural selection had more generally fallen out of favour.[28]

Diluting Darwinism

When we look more closely at competing mechanisms for evolution a different problem arises. It was addressed by James Moore in his book *The Post-Darwinian Controversies* (1979). Is it appropriate to use the label "Christian Darwinians" for scholars who, unlike Poulton and Dixey, did not grant primacy to natural selection? Moore elaborated a complex taxonomy to differentiate between, on the one hand, Christian evolutionists who developed rather facile, liberal evolutionary systems, often based on Lamarckian ideas and optimistic assumptions about evolutionary 'progress', and, on the other, more conservative Christians who were often the ones to take seriously and appreciate fully the challenge of natural

---

[26] Cited by Arthur Peacocke, "Biological Evolution and Christian Theology – Yesterday and Today," in Durant, *Darwinism and Divinity*, 101-30, 111.

[27] Brooke, "Darwin's Science and his Religion," 46-7.

[28] Richard England, "Natural Selection, Teleology, and the Logos: From Darwin to the Oxford Neo-Darwinians," *Osiris* 16 (2001): 270-87; Peter J. Bowler, *The Eclipse of Darwinism* (Baltimore: Johns Hopkins University Press, 1983).

selection. It was they, Moore controversially argued, who also had the best resources for assimilating the theory that Darwin had actually propounded. There is clearly a need for some distinction of this kind, given the perennial temptation for theologians to select from the sciences only those features that suit their case. In his book *Deeper than Darwin,* John Haught observes that it is still the case that when theologians talk about evolution at all they typically edit out the most challenging features of the Darwinian and neo-Darwinian pictures of life.[29]

There are, however, difficulties with too filigree a taxonomy because Darwin himself reduced the emphasis he gave to natural selection between the first and sixth editions of his *Origin of Species.* In his *Descent of Man* (1871), he owned to having exaggerated its importance in relation to other considerations, such as the role of environmental pressures and the effects of the use and disuse of bodily parts. This makes it difficult to give a precise definition of a 'Darwinian' that is not anachronistic. What can be said is that among Christians receptive to the reality of an animal ancestry for humankind, there was a common, if not universal, tendency to modify Darwin's mechanism in ways that seemed more compatible with belief in a guiding providence.[30] A classic example would be Henry Drummond's insistence that to the Darwinian struggle for life had to be added another principle – the struggle for the life of others, as in parental care for offspring.[31] Peter Bowler has noted the irony that, before Darwin, the evolutionary theory of Jean-Baptiste Lamarck was frequently vilified for its affirmation of continuity between humans and animals; whereas, after Darwin, when the case for biological evolution was so much stronger, Lamarckian mechanisms for evolution proved rather attractive to religious apologists troubled by the seemingly undirected agency of natural selection. Less troubling was the idea that there were propensities inherent within living organisms that had putatively driven, and given direction to, evolutionary development.[32] Some Christian commentators, notably the Catholic evolutionist, St George Mivart, were deeply critical of natural selection and of Darwin's emphasis on the gradual accumulation of small

[29] John F. Haught, *Deeper than Darwin: The Prospect for Religion in the Age of Evolution* (Boulder, Colorado: Westview Press, 2003).
[30] Frederick Gregory, "The Impact of Darwinian Evolution on Protestant Theology in the Nineteenth Century," in *God and Nature: Historical Essays on the Encounter between Christianity and Science*, ed. David C. Lindberg and Ronald L. Numbers (Berkeley: University of California Press, 1986), 369-90.
[31] Henry Drummond, *The Ascent of Man* (London: Hodder and Stoughton, 1904), 15-47.
[32] Bowler, *Eclipse of Darwinism*, 8-9 and 50-66.

variations. Mivart's critique, published in his *Genesis of Species* (1871), offended Darwin's inner circle who considered it religiously biased and unfair. Darwin himself was incensed because whereas he had tried to give a balanced account of the pros and cons of his theory, Mivart appeared to focus only on the objections.[33] There certainly were limits beyond which a believer in evolution could be judged not to be a Darwinian.

Mivart illustrates yet another divisive issue – whether, as a Christian, one could accept that the emergence of the human soul was a 'natural' consequence of evolution, or whether this would be to concede too much. Mivart allowed evolution to explain the derivation of the human body, but the soul remained a special addition, a divine gift at a critical juncture in hominid development. To this day, his dualistic model of body and soul has featured prominently in Catholic discussions of evolution. For Darwin himself such a response would smack of compromise. But there was, and is, a difficulty here in going the whole way with the Darwinians. This is because Darwin's philosophy of mind, as early as the first sketches of his theory in the late 1830s, really was monistic rather than dualistic. Much later, after his *Descent of Man* was published, to be a Christian *Darwinian* was to come face to face with a naturalistic account of the origins of religion itself.[34] There is a passage in *The Descent of Man* that presages the thinking of those cognitive psychologists today who emphasise the selection of *hyper*sensitive agency detection systems in the early religious history of humankind.[35] Darwin recalls how one summer day, when a parasol was swaying in the breeze, his dog began to bark, apparently surmising that there was an invisible intruder in his territory. For Darwin there was an obvious parallel between the behaviour and thought processes of the dog and early human animism.[36] One is tempted to say that his proto-religious dog began to illuminate a whole train of theological *dogmatics*! Perhaps not; but to be told why, ultimately, one believes what one believes can be an uncomfortable experience! Aesthetically, it remains offensive to the majority of Darwinian naturalists to have to make an exception of the human soul.

---

[33] Brooke, '"Laws impressed on matter"?', 271-2; Brooke and Cantor, *Reconstructing Nature*, 255-62.

[34] Matthew Day, "Godless Savages and Superstitious Dogs: Charles Darwin, Imperial Ethnography and the Problem of Human Uniqueness," *Journal of the History of Ideas* 69 (2008): 49-70.

[35] Justin L. Barrett, *Why Would Anyone Believe in God?* (Lanham MD: AltaMira Press, 2004).

[36] Charles Darwin, *The Descent of Man* (London: Murray, [1871] 1906), 144-5.

Darwinism and theodicy

Darwin's eventual resistance to a Christian theism provides an indication of another issue that was crucial to his defection and which proved divisive for Christians actively seeking to construct a theodicy. This was the sheer extent of pain and suffering in nature. For Darwin, this spoke against belief in a beneficent deity but was comprehensible on his theory of natural selection. As Darwin pointed out, the theological problem was exacerbated by the fact that the suffering of animals could not be explained away as a vehicle of their moral improvement.[37] Rather belatedly perhaps, the problem of animal suffering has come to the fore in recent discussions of theodicy.[38] In this context Christopher Southgate has argued that there can be no solution to the problem unless animals are granted their own immortality, a view that he also finds in John Wesley and Keith Ward.[39] But here is a perfect example of the kind of divisiveness that serious engagement with Darwinism tends to produce. As Southgate has acknowledged, of all the theses in his book, the postulate of a heaven populated by every species is the one that has proved most controversial for his audiences. The divisiveness goes deeper still when he notes that even for those sympathetic to the idea, there is an issue whether all creatures are to find their place in heaven or whether only representatives of their species will enjoy redemption. Would not one mosquito suffice?

Remaining with questions of theodicy, it is striking how Darwinism has been seen as a resource for rationalising pain and suffering in the world. If a competitive struggle for existence and the operation of natural selection constitute the motor of evolution, pain and suffering might be deemed the price for a process without which no morally responsible creatures would have emerged. This was how Asa Gray argued for the compatibility of natural selection with an evolutionary teleology:

> Darwinian teleology has the special advantage of accounting for the imperfections and failures as well as for successes. It not only accounts for

---

[37] Nora Barlow, ed., *The Autobiography of Charles Darwin* (London: Collins, 1958), 90.

[38] See, for example, Christopher Southgate, *The Groaning of Creation: God, Evolution and the Problem of Evil* (Louisville: Westminster John Knox Press, 2008); Michael J. Murray, *Nature Red in Tooth and Claw: Theism and the Problem of Animal Suffering* (Oxford: Oxford University Press, 2008); Celia Deane Drummond and David Clough, eds., *Creaturely Theology: On God, Humans and Other Animals* (London: SCM Press, 2009).

[39] Southgate, *The Groaning of Creation*, 78-82.

> them, but turns them to practical account. It explains the seeming waste as being part and parcel of a great economical process. Without the competing multitude, no struggle for life; and without this, no natural selection and survival of the fittest, no continuous adaptation to changing surroundings, no diversification and improvement leading from lower up to higher and nobler forms. So the most puzzling things of all to the old-school teleologists are the *principia* of the Darwinian.[40]

A not dissimilar argument is still to be found in texts on Christianity and evolution, as in Francisco Ayala's recent book *Darwin's Gift to Science and Religion*.[41] But this style of apologia can itself be divisive because many have preferred to follow Immanuel Kant, who in his reflections on the book of Job, declared the failure of any philosophical theodicy.[42] There are still other difficulties that advocates of a Darwinian theodicy have to face – principally how can we know that no other mode of creativity was possible for an omnipotent and merciful Creator? As Darwin's disciple in the study of animal behaviour, George Romanes, observed, one of the greatest problems Darwinism posed for theism was that the nature of the deity inferred on the basis of the theory was at variance with the attributes presupposed in the noblest form of religion:

> If the general order of Nature is due to Mind, the character of that Mind is such as it is conceived to be by the most highly developed form of religion. A conclusion which is no doubt the opposite of that which we reached by contemplating the phenomena of biology; and a contradiction which can only be overcome by supposing, either that Nature conceals God, while man reveals Him, or that Nature reveals God while man misrepresents Him.[43]

Could nature be said to reveal God if it could be shown that the Darwinian mechanisms were the only viable programme for producing human beings? But, again, the question is how could this be known? Southgate's discussion of theodicy illustrates the problem. Addressing the sceptic's question, "Why did God not just create heaven?" Southgate answers as follows:

---

[40] Gray, *Darwiniana*, 310-11.

[41] Francisco Ayala, *Darwin's Gift to Science and Religion* (Washington: Joseph Henry Press, 2007).

[42] Kant's *On the Failure of All Attempted Philosophical Theodicies* (1791) is discussed in Brooke, *Science and Religion*, 203-9.

[43] George Romanes, *Thoughts on Religion* (London: Longmans, Green and Co., 1896), 83.

> Since this was the world the God of all creativity and compassion chose for the creation of creatures, we must presume that this was the only type of world that would do for that process.

The word "only" is doing a lot of work here even when it is acknowledged that theology necessarily deals in presumption not proof. We are ultimately led to what Southgate concedes is a guess: "Our guess must be that though heaven can eternally preserve those [creaturely] selves, subsisting in suffering-free relationship, it could not give rise to them in the first place."[44] The problem, divisive and accentuated by Darwin, is whether this might be seen as placing a speculative constraint on omnipotence.

Providence and progress

Back in the nineteenth century, Asa Gray's defence of Darwin reveals yet another divisive issue. This was whether to make theological capital from what Darwin's theory was as yet unable to explain. The temptation to opt for a god-of-the-gaps has repeatedly recurred in Christian engagement with the sciences. Darwin had not explained how life first arose; nor could he explain how the variations on which natural selection worked came about. In these circumstances it was tempting for Gray to propose a role for the deity in micro-managing the variations, ensuring they were propitious for progress. It was with Darwin's resistance to this notion that this essay began. The dangers for Christian theology in Gray's proposal would become all too clear once a genetic mechanism was found for the source of variation. The closure of gaps as the sciences have advanced is a theme with its own variations in the post Darwinian debates, notably in the infilling of the supposedly missing links in the fossil record. But the general issue here has been a perennial source of dissension among Christian Darwinians, and indeed Christian anti-Darwinians, as the current preoccupation with 'intelligent design' makes clear.[45]

Another tempting but divisive strategy has been to incorporate the idea of progress or directionality into models of theistic evolution. Claims for an overarching progression towards creatures with human capacities have featured in many theologies of nature; but there are well known problems here. In the absence of anthropocentric assumptions it is not clear that

---

[44] Southgate, *The Groaning of Creation*, 90.

[45] William A. Dembski and Michael Ruse, eds., *Debating Design: From Darwin to DNA* (Cambridge: Cambridge University Press, 2004); Michael Ruse, *Darwin and Design* (Cambridge: Harvard University Press, 2003), 313-36.

merely from knowledge of evolutionary processes any compelling directionality can be inferred – certainly not a simple linearity. The assumption of progress *towards* humans tends to be axiological. It is as if the old chain of being is still superimposed on Darwin's branching tree, even though the systems are barely, if at all, superposable. And there is still the further divisive issue. Is human uniqueness to be established on the basis of biological and psychological criteria that make a covenant between humans and God somehow natural and comprehensible? Or is human uniqueness to be established theologically, independently of biological criteria and only on the basis of a certified interest in humankind taken by the deity? Putting the question more sharply, is it necessary to be able to specify uniquely human capacities before it becomes plausible to claim a divine concern for humanity?

The difficulty of reducing evolutionary history to a single overarching divine purpose was already clear to the Catholic modernist George Tyrrell in the early years of the twentieth century. Tragedies like the earthquake that hit Southern Italy in 1908 could not be slotted into a single all-embracing picture. Tyrrell rejected that idea in favour of interpreting nature as a far more complex work of art with many different aims and meanings. The image of the branching tree was even made to do theological work. As he put it:

> [The universe] teems with aims and meaning, although it has no one aim or meaning. It is like a great tree, that pushes out its branches, however and wherever it can, seeking to realize its whole nature … in every one of them, but aiming at no collective effort. This is its play, this is its life, this is, if you will, its end.[46]

Darwin's own image of the branching tree, as Gillian Beer has observed, is itself deeply ambiguous.[47] On the one hand it confutes a simple directionality; on the other, the growth of a tree is, nevertheless, upwards. So I turn now to three Christian Darwinians who reflected deeply on what "upwards" might mean for the future of the universe and, of course, humankind. In their different ways, each made connections between providence and progress. Each held theological views that in one context or another were controversial. I refer to Pierre Teilhard de Chardin, Ronald Fisher, and Theodosius Dobzhansky – a Catholic, an Anglican and a member of the Eastern Orthodox tradition.

---

[46] Brooke and Cantor, *Reconstructing Nature*, 163-4.

[47] Gillian Beer, *Darwin's Plots: Evolutionary Narrative in Darwin, George Eliot and Nineteenth-Century Fiction* (London: Routledge and Kegan Paul, 1983).

Teilhard told a story of increasing complexification in the history of life that had eventuated in the emergence of consciousness. The history was as yet incomplete because it had still to converge at an 'omega point' when humanity would achieve those qualities and capacities prefigured in the life of Christ.[48] In his spiritual writings Teilhard challenged Christians to think of themselves as collaborators with the deity in perfecting the world.[49] This was an arresting but divisive image, appearing too presumptuous to many. But another of Teilhard's doctrines, his teaching on original sin, proved even more divisive and may have contributed to his exile in China. In one of his early notes, he wrote:

> Original sin, taken in its widest sense, is not a malady specific to the Earth, nor is it bound up with human generation. It simply symbolizes the inevitable chance of evil … which accompanies the existence of all participated being.[50]

The emphasis, in Teilhard, of a world still to be completed appeared in many other twentieth-century theological constructs. The influential geneticist Ronald Fisher was an Anglican who considered it a religious duty to ensure that evolutionary progress was not compromised by social trends that he considered injurious. The idea that humans have a unique responsibility to preside over and manage the future course of evolution is one that we correctly associate with the secular evolutionary humanism of Julian Huxley and latter day prophets of post-human capacities. But, as recent research has revealed, in both Britain and America Christians were sometimes seriously committed to eugenic agendas. Fisher is an example. He believed there was a religious duty to steer the human race upwards and away from degeneration. In common with the Bishop of Birmingham, Ernest Barnes, Fisher attended meetings of the Eugenics Society. In concert with Leonard Darwin, Fisher believed that eugenics was the common ground on which religion and science should unite to improve the race. Positive eugenic policies were required to raise the birth rate of the ablest. As James Moore has recently put it, humans to Fisher were not like other animals; they inevitably degenerate, superior classes are swamped, civilizations decay: "Positive eugenics was the sole scientific means of

---

[48] Pierre Teilhard de Chardin, *The Human Phenomenon*, trans. Sarah Appleton-Weber (Portland: Sussex Academic Press, 1999).

[49] Pierre Teilhard de Chardin, *Le Milieu Divin* (London: Collins, 1960), 49-73.

[50] Cited by Haught, *Deeper than Darwin*, 166-7.

moving humanity onwards and upwards to the Divine Goal."[51] Fisher's own proposal was that there should be differential family allowances, the larger allowances paid to the *higher* socio-economic groups. The degree to which Christian Darwinians were actively involved in promoting eugenic policies comes as an unwelcome surprise to many today. Helping the tree to grow upwards has been a deeply divisive enterprise.[52]

For my third example I turn to Dobzhansky because one of his aphorisms has achieved canonical status, particularly in anti-creationist polemics: "Nothing in biology makes sense except in the light of evolution." One finds that declamation all over the 'world wide web'. What is less well known, clearly to many who reproduce that motto, is that Dobzhansky remained faithful to his Eastern Orthodox Christianity, seeing no clash with a science of genetics and Darwinian evolution.[53] Indeed as an admirer of Paul Tillich, Dobzhansky saw a profound compatibility between Christianity and evolutionary biology, as long as the salient biblical texts were interpreted existentially and symbolically. Hence the following remarks in his book *Mankind Evolving*:

> The meaning of the acquisition of self-awareness in human evolution is expressed beautifully in the biblical symbol of the Fall of Man. Self-awareness is a blessing and a curse. Through self-awareness man attained the status of a person in the existential sense: he became conscious of himself and of his environment. … Self-awareness and foresight brought, however, the awesome gifts of freedom and responsibility. Man knows that he is accountable for his acts: he has acquired the knowledge of good and evil. This is a dreadfully heavy load to carry. No other animal has to withstand anything like it. There is a tragic discord in the soul of man.[54]

Dobzhansky's existentialist interpretation of a biblical story does, however, underscore another division. This opened up during the nineteenth

---

[51] James Moore, "Ronald Aylmer Fisher: A Faith Fit for Eugenics," in Nicolaas A. Rupke, *Eminent Lives in Twentieth-Century Science and Religion* (Frankfurt: Peter Lang, 2007), 103-38, 126.

[52] On eugenics within the churches, see Peter J. Bowler, *Reconciling Science and Religion: The Debate in Early Twentieth Century Britain* (Chicago: University of Chicago Press, 2001), especially 260-70; and Christine Rosen, *Preaching Eugenics: Religious Leaders and the American Eugenics Movement* (Oxford: Oxford University Press, 2004).

[53] Jitse Van Der Meer, "Theodosius Dobzhansky: Nothing in Evolution Makes Sense Except in the Light of Religion", in Rupke, *Eminent Lives*, 79-101.

[54] Theodosius Dobzhansky, *Mankind Evolving* (New Haven: Yale University Press, 1962), 338.

century and continues to separate many liberal from more conservative Christians. This is the degree to which any *detailed* harmonisation should be sought between sacred texts and the sciences. Darwin embarrassed, but did not terminate, the harmonisation programme that had developed during the first half of the nineteenth century when geologists were reshaping Earth history.[55] William Gladstone is an example of a Victorian Christian scholar who continued to harmonise *Genesis* with scientific cosmologies and was molested by Huxley as a consequence.[56] The issue is still not dead. How often has one heard the argument that the *Genesis* narrative is compatible with evolutionary theory because it asserts that creation took place in stages and that man had his origins in the dust of the earth?

Chance, omnipotence and kenosis

I referred earlier to Darwin's tentative formula that biodiversity is the result of designed laws with the details left to 'chance'. This is a useful reminder that Christian Darwinians have had to deal with another problem: what construction is to be placed on 'chance' in the context of evolutionary mechanisms. For some the very word 'chance' has been construed in antithesis to design, as it was by the Princeton theologian Charles Hodge. In his influential book *What is Darwinism?*, Hodge did not accuse Darwin of atheism. Nor did he regard the idea of evolution as intrinsically atheistic. But the Darwinian theory, with its ateleological mode of explanation, was for Hodge destructive of design and theologically unacceptable.[57] By contrast there have been Christian scholars eager to protest that elements of randomness in evolutionary processes should not be seen as destructive of providentialist interpretation.[58]

By a 'chance' event Darwin emphatically did not mean one that was uncaused, but one that rather resulted from the intersection of otherwise

---

[55] Brooke, "Genesis and the Scientists".

[56] Thomas H. Huxley, "Mr. Gladstone and Genesis" [1886], in *Science and Hebrew Tradition: Essays by Thomas H. Huxley* (London: Macmillan, 1904), 164-200.

[57] David N. Livingstone, *Darwin's Forgotten Defenders: The Encounter Between Evangelical Theology and Evolutionary Thought* (Grand Rapids: Eerdmans, 1987), 102-5.

[58] See, for example, David Bartholomew, *God, Chance and Purpose* (Cambridge: Cambridge University Press, 2008) and Paul Ewart, "The Necessity of Chance: Randomness, Purpose and the Sovereignty of God," *Science and Christian Belief* 21 (2009): 111-131.

independent causal chains. The real issue for him, however, concerned the randomness in the appearance of variations, many of which were deleterious to the organism in question. As we saw earlier, Darwin differed from Gray because he could not accept that variations were produced with a *prospective* use in mind. But what follows for a theology of nature if the potential for novelty is introduced by seemingly accidental or random events? The division here is between those who like to believe that there is a detailed blueprint for creation, to which the evolution of nature adheres, and those who prefer to associate creativity with spontaneity. In his reply to Jacques Monod's nihilistic conclusions on chance and necessity, Arthur Peacocke saw in the interplay between the randomness of variation and the rigour of natural selection an exquisite mechanism for the exploratory development and actualisation of a God-given potential built into nature.[59] Christian Darwinians of earlier generations were not always fazed by the problem. A former President of Princeton, James McCosh, simply observed that the prevalence of accident in evolution cannot be accidental.[60] His point was that God governs the world not only by laws but by a "complication" of laws that produces "fortuities".[61] For the American pragmatist, Charles Peirce, there was no difficulty in accepting absolute chance in nature. The execution of predetermined ends would be a purely mechanical process, leaving no room for novelty and development. But if there was a personality behind the universe, diversification, spontaneity and potential for growth were precisely what one would expect.[62] A tension still exists between Christian commentators who would like to conserve a sense of the micro-management of evolution and those, such as John Haught, who warn that the doctrine of Providence is not about God's coercive control, perfect design, or blueprint for progress. Divine care, according to Haught, means the opening of the world to an always new future. Providence means promise.[63] Far from being a threat, the science of evolution becomes, again, a friend to Christianity because it helps to recover a truly biblical spirit of hope. In Haught's own words:

---

[59] Arthur Peacocke, *Creation and the World of Science: The Re-Shaping of Belief* (Oxford: Oxford University Press, [1979] 2004), 5-52, 103-11.
[60] Moore, *The Post-Darwinian Controversies*, 245-50.
[61] Moore, *The Post-Darwinian Controversies*, 335.
[62] Brooke, *Science and Religion*, 315-6.
[63] Haught, *Deeper than Darwin*, 40, 53, 63-5, 143-4.

> The new evolutionary accounts of nature invite us to recapture the often obscured portrait of a self-humbling, suffering God who is anything but a divine controller or designer of the cosmos.[64]

Haught's sophisticated discussion is attractive in many ways. At the same time, it exposes the division between theologies fertilised by the process philosophy of Whitehead and the anthropomorphism of an older natural theology in which God was reduced to a magician, manipulator or craftsman – and where the deity still remains in the minds of many.

This leads us to a deeper division still. It relates to one of the more enduring aspects of Darwin's impact on Christian theology. I am thinking of the reconsideration of what it means to speak of God's omnipotence. Theologians appropriating the process philosophy of Whitehead have often seen in biological evolution an additional justification for replacing a coercive managerial God with the more modest persuasive agency of a demiurge.[65] This is the divine lure that participates in and draws the world into the future rather than pushing it, as it were, from behind. On this view to speak of God's omnipotence is a theological mistake, as it was for Charles Hartshorne.[66] But there have been other ways of modelling God's activity in the world. In these the limitations of God's power are not intrinsic to the nature of God but are self-imposed, self-restraining, self-humbling. The analogy is with kenotic understandings of God's self-emptying in Christ.[67] The two approaches, the Platonist and the kenotic would seem to be quite different, though attempts are certainly made to take the best from both. Haught, for example, appeals to Whitehead in many places. But he also speaks of a God who "refrains" from wielding a domineering kind of power.[68] This of course invites the further question, whether there might not be exceptional circumstances in which God might be expected to refrain from that self-restraint if by so doing a great evil

---

[64] Haught, *Deeper than Darwin*, 81.

[65] For an introduction to Whitehead's doctrine of God, see John B. Cobb Jr, *A Christian Natural Theology*, 2nd edition (Louisville: Westminster John Knox Press, 2007), 81-111.

[66] Charles Hartshorne, *Omnipotence and Other Theological Mistakes* (New York: State University of New York, 1984).

[67] John Polkinghorne, "Kenotic Creation and Divine Action", in *The Work of Love*, ed. John Polkinghorne (Grand Rapids: Eerdmans, 2001), 90-106; Peter Colyer, "Kenosis by Addition, Kenosis by Subtraction," in *Humanity, The World and God*, Studies in Science and Theology, vol. 11, ed. Willem B. Drees, Hubert Meisinger and Taede A. Smedes (Lund: University of Lund, 2007-8), 69-82.

[68] Haught, *Deeper than Darwin*, 82.

might be countermanded. Does the self-imposed restraint have to be thought of as irreversible?

Conclusions

What conclusions can be drawn from this survey of divisiveness? On the one hand, a house divided against itself cannot stand. If that is so, then Darwin's impact on Christianity has been destructive, albeit in a subtler way than is popularly supposed. On the other hand, controversy has been the lifeblood of theology, as it has for science. The story I have briefly told can be read as a summary of sustained attempts to reconstruct a theology of nature that takes seriously the science of evolutionary biology and how it has itself evolved. It is difficult to deny, however, that many of these theologies of nature are primarily efforts to achieve *compatibility* with neo-Darwinian perspectives – a more modest venture than was enshrined in the classic natural theology literature. Even where claims are made for a revival of natural theology, as in Alister McGrath's recent work, the apologia is not free-standing but is explicitly subsumed within a pre-existing framework of Christian theology.[69] In that respect Asa Gray's divergence from Darwin, with which I began, does epitomise what was to become a general shift. The initiative was wrested from science-based arguments for God's existence and largely replaced by attempts to achieve conciliation. The extent to which the survival of the Christian religion has depended on such conciliation is a very different question, given the disconcerting success of anti-Darwinian forms of fundamentalism. In that connection it can be salutary to remember an observation of William James. As a Darwinian student of religious experience, James had this to say about the survival of the religious:

> Every sort of energy and endurance, of courage and capacity for handling life's evils, is set free in those who have religious faith. For this reason the strenuous type of character will on the battle-field of human history always outwear the easy-going type, and religion will drive irreligion to the wall.[70]

In a Darwinian universe, it was the strenuously religious who were best fitted to survive.

---

[69] Alister McGrath, *The Open Secret: A New Vision for Natural Theology* (Oxford: Blackwell, 2008), 171-217.
[70] William James, *The Will to Believe and Other Essays in Popular Philosophy* (New York: Longmans, Green and Co., 1899), 213.

CHAPTER FIVE

RESPONSE: FURTHER VARIETIES OF CHRISTIAN DARWINIAN

DAVID KNIGHT

It is always a pleasure to follow John Brooke as he applies fine taxonomic and contextual skills to unpick the complexities of past intellectual life. We are all too prone to generalise about decades and to corral our ancestors, when we know (and it is one of the advantages of ageing) how the prating and pontificating of journalists and politicians fail to fit our own experience of the 1950s or '60s. While we cannot and should not avoid talk of nations and generations, John is very good at bringing out particularity and individuality; what we thought we knew looks fuzzier as we relinquish the benefit of hindsight and see real people struggling to make sense of their lives and their communities, and changing their minds in the light of their experiences. I hope to do a little more of the same in drawing attention to some people, and the groups to which they belonged with more or less fervour, trying to avoid categorising them in terms they would not have comprehended.

John highlights reactions to the shock from the middle years of the nineteenth century, conveniently identified with 'Darwin' and the *Origin*, 1859. He notes how odd (and alarming) it was to realise that natural sciences don't necessarily support religion; that cruelty and 'chance' rather than design seemed to lie behind natural selection; that old distinctions like bodies/souls, immanence/transcendence, were problematic; and that if one accepted evolution and open-endedness (and with it, kenosis), the doctrine of God's omnipotence seemed compromised. Materialism, which Priestley had tried to Christianise, remained a terrible bugaboo, taken to lead inexorably to atheism and immorality. Being a Christian Darwinian was not straightforward; nevertheless, as we know from John, many in Thomas Huxley's generation managed it in different degrees and in different ways. Here are a few more, adding to John's crowd of witnesses, from those years after Robert Chambers' sensational *Vestiges,* 1844, and Alfred Tennyson's *In Memoriam,* 1850, which we can for convenience label 'Darwinian'. We can group them loosely as parson naturalists, Christian agnostics, Incarnationalists, and pillars of the Establishment.

Parson naturalists

The 'parson-naturalist' now definitely belongs to an endangered species, but that was the role to which Darwin and some at least of his Oxbridge contemporaries aspired when as a young ordinand he embarked on *HMS Beagle,* and was a significant group in the nineteenth century.[1] The pattern went back to John Ray in seventeenth-century Cambridge, whose *Wisdom of God Manifested,* 1691, is a classic, and after whom an important natural-history publishing society founded in the nineteenth century is named. Many played an important role in natural history. They were aware how perilously close William Paley came in his *Natural Theology,* 1802, to congratulating God on a particularly ingenious contrivance;[2] and generally preferred a theology of nature over 'evidences'. Thus William Symonds, Rector of Pendock, Justice of the Peace and Lord of the Manor, was a Fellow of the Geological Society and a member of its Council, and published 43 papers. He was a founder, and for many years President, of the Malvern Field Club. His daughter, the widow of the naturalist Sir William Jardine, became the second wife of Joseph Hooker, Director of Kew and one of Darwin's closest friends. Symonds was unfazed and generally irenic in his response to the *Origin,* concluding his *Old Bones,* 1861 edition:[3]

> The examination of the principles of Palaeontology and Biology has convinced [natural] philosophers that there is a 'constantly operating, secondary, creational law'.

He was happy to associate Richard Owen and Huxley with Darwin in this conclusion, which promised to make natural history a science based on law, like astronomy, rather than simply upon taxonomy. With him,

---

[1] Nicholas Cooper, ed., *John Ray and his Successors: the Clergyman as Biologist* (Braintree: John Ray Trust, 2000).

[2] William Paley, *Natural Theology* [1802], ed. Matthew Eddy and David Knight (Oxford: Oxford University Press, 2006).

[3] William S. Symonds, *Old Bones, or Notes for Young Naturalists*, 2nd ed. (London: Hardwicke, 1862) 120.

going a long way with his friend Darwin, was Leonard Jenyns, Vicar of Swaffham Bulbeck, who had been first choice to sail on *HMS Beagle,* but (being at the outset of his career) declined the opportunity. He published works of natural history, and founded a natural history field club in Bath, to which he retired in 1860. Canon Henry Tristram of Durham, an authority on the natural history of the Holy Land (a curious bird common there, 'Tristram's grackle', is named after him) wavered between full and partial conviction. Not only Tristram, but also Huxley, differed from Darwin over the place of natural selection in evolution: Darwinism, like religious experience, had its varieties. But if we look at practice, rather than at dogma, we discover that parson-naturalists found themselves willy-nilly taking some kind of evolutionary perspective, probably more optimistic than Darwin's, as time went by. They were well-aware of the 'deep time' that geologists had been taking for granted since the Geological Society's foundation in 1807, popularised not only by Charles Lyell but also by Adam Sedgwick and William Buckland, clergymen and professors of Geology at Cambridge and Oxford.[4] Thus they did not need to share in the panic (stoked by Samuel Wilberforce) associated with *Essays & Reviews,* 1860, about new ideas (including 'Darwinism') coming into old-time religion; or the alarm about Bishop Colenso's foray into *kenosis* and arithmetic that provoked the summoning of the first Lambeth Conference, 1867 (episcopal rows and anathemas are not novelties).

Christian agnostics

Parson-naturalists were united in their *practice*, but our next group whom we might call 'Christian agnostics' (though some were post-Christians) shared *ideas*. The Oxford logician Henry Mansel delivered a famous series of *Bampton Lectures,* 1858, in which he promoted the classic *via negativa.*[5] Any propositions about God must be false; we can safely say only that He is neither this nor that. This way of thinking was not much taken up by Victorian churchmen, though Mansel became Dean of St Paul's in 1868; but it chimed in with Herbert Spencer and Huxley's

---

[4] Cherry Lewis and Simon Knell, eds., *The Making of the Geological Society of London* (London: Geological Society, 2009); Martin J.S.Rudwick, *Bursting the Limits of Time: the Reconstruction of Geohistory in the Age of Revolution* (Chicago: University of Chicago Press, 2005).

[5] Henry L. Mansel, *The Limits of Religious Thought*, 4th ed., (London: John Murray, 1859).

'New Reformation' and the ideas of their X-club.[6] Huxley coined the word 'agnostic' for those sharing his refusal to dogmatise about what could not be proved, taking care to separate himself from Comtean Positivists and their Religion of Humanity, and from atheists (taken in Victorian Britain to be libertines): he believed that this was the true spirit of science. John Stuart Mill was indignant that Mansel seemed to be stealing the clothes of honest agnostics while drawing a salary as a clergyman; and others felt the same way. Nevertheless, these 'scientific naturalists', with their Sunday lecture Society and their Biblical rhetoric, were more anti-clerical than anti-Christian. John Tyndall's notorious Presidential Address to the British Association for the Advancement of Science in Belfast in 1874, when he claimed that science would "wrest from theology" cosmology and the cultural high ground, was not 'scientistic': music, literature, fine art, and individual religion, like his own pantheistic reflections on mountain tops, were highly important and distinct from science.[7] By the end of his life, and to the horror of Spencerians, Huxley in his *Romanes Lectures,* 1893, suggested that civilised ethics required us to oppose natural selection, helping women and children into the lifeboat first. Battle lines were fuzzier than they might appear, as we can see also in Alfred Russel Wallace, the 'co-discoverer', whose letter from Malaysia set Darwin writing the *Origin* (an abstract, without notes or bibliography, of a three-volume work on *Natural Selection* that never appeared). Wallace diverged from the 'mainstream' Darwinians, taking a different line in emphasising human distinctness and in embracing spiritualism, along with socialism, and in spirited opposition to vaccination.[8] Darwinism was a broad church.

Incarnationalists

More strictly Christian Darwinians were those we can call 'Incarnationalists' because they emphasised Incarnation rather than Atonement as central to Christianity. They came from very diverse traditions. Henry Drummond, trained at New College Edinburgh for the

---

[6] Mark Francis, *Herbert Spencer and the Invention of Modern Life* (Stocksfield: Acumen, 2007).

[7] Bernard Lightman, "Scientists as Materialists in the Periodical Press: Tyndall's Belfast Address," in *Science Serialized: Representations of Science in Nineteenth-century Periodicals*, ed. Geoffrey Cantor and Sally Shuttleworth (Cambridge MA; MIT Press, 2004), 199-237.

[8] Alfred Russel Wallace, *My Life*, new ed. (London: Chapman & Hall, 1908); Martin Fichman, *An Elusive Victorian: the Evolution of Alfred Russel Wallace* (Chicago: University of Chicago Press, 2004).

Free Kirk ministry, and associated with missions in Africa, London and Edinburgh, was an enthusiast for altruism and progress rather than gloomy emphasis upon the Fall and human sinfulness. In 1894 his Lowell Lectures delivered in Boston, Massachusetts, were published as *The Ascent of Man,* deliberately echoing Darwin's *Descent of Man,* 1871, but emphasising development, and in Lamarckian vein how 'getting on' was the mainspring of progress in the evolution of animals and of human society: egoism plus altruism made the world go round, and brought morality into it. Further Evolution must go on, and the Higher Kingdom come; we are not at the culmination of the process.[9]

John Henry Newman had much earlier preached and written about the development of doctrine, as he moved towards the Roman Catholic Church in 1845. John Richardson Illingworth came from the Anglican high-church tradition revived in Oxford in the previous decade by Newman and the other Tractarians, and was the moving spirit of the group which published, under Charles Gore's editorship, *Lux Mundi,* 1889, a volume of essays that turned out to be much less provocative than *Essays & Reviews,* by an earlier Oxonian team, but astonishingly popular. Aubrey Moore's paper on the Christian Doctrine of God has been mentioned by others; but Illingworth himself contributed two essays, one dealing specifically with the Incarnation and development.[10] He welcomed evolutionary development because it meant closing the gap between the sacred and the profane; and made the world an organism, rather than the great clockwork imagined by Deists that God had wound up and from which He then stood aside. Creation was an ongoing process, not an event. In 1894, Illingworth delivered a set of Bampton Lectures. For people in these Incarnationalist traditions, theology was not coming to terms with science by surrendering territory for peace: rather, science was helping to widen and deepen theologians' vision of God.

Pillars of the Establishment

As a fourth category, we look at 'pillars of the Establishment'. Frederick Temple, then Headmaster of Rugby, had contributed to *Essays and Reviews,* and shared in the odium poured by the supposedly great and the good upon their heads as the 'Seven against Christianity'. Nevertheless

---

[9] Henry Drummond, *The Lowell Lectures on the Ascent of Man* (London: Hodder and Stoughton, 1894).

[10] John R. Illingworth, "The Incarnation in Relation to Development," in *Lux Mundi: a Series of Studies in the Religion of the Incarnation*, 15th edition, ed. Charles Gore (London: Murray, [1889] 1899), 132-57.

in 1869 William Gladstone as Prime Minister had appointed him Bishop of Exeter, despite his refusal to withdraw his essay (it was an inoffensive piece on the education of the world, obnoxious only on account of the company it was keeping). Having warned the publishers so that they could print a substantial run of the existing edition, once in office he declared that he could not now as a Bishop sanction his essay appearing in any future editions of a possibly heterodox book, and in due course it went out of print.[11] He remained rather anti-clerical, even as he became increasingly respectable, becoming successively Bishop of London (1885), and Archbishop of Canterbury (1897). I happen to have Bishop Hensley Henson's copy of Temple's biography, and he has marked a passage in a letter where (admittedly in a quite different context) Temple approvingly quoted Chrysostom: "I verily believe that very few clergymen will be saved."[12] In 1884, Temple delivered a set of Bampton Lectures on the relations between science and religion in which (as John Brooke and others have remarked) drawing particularly upon Kant and S.T. Coleridge, and in the spirit of *Essays and Reviews* but perhaps with greater gravitas, he welcomed evolutionary theory. Naturally, because it does not make good copy, less is made of this in the wider world than of Wilberforce's earlier denunciations.

One could hardly be more of a pillar of the Establishment than Arthur Balfour. His uncle and patron, Lord Salisbury, was the gloomy aristocrat who as Prime Minister had led Britain into the twentieth century; and Balfour was his successor. Like the Queen an Anglican in England and a Presbyterian in Scotland, he approved of the revival of Congregation and took church appointments and church views very seriously. He was extraordinarily well-connected by family ties; and his in-laws included the philosopher and reluctant agnostic Henry Sidgwick, founding President of the Society for Psychical Research; the physicist and Cavendish Professor at Cambridge Lord Rayleigh; and Edward Benson, Temple's friend and predecessor as Archbishop of Canterbury. A gentleman-commoner at Trinity College, Cambridge, tutored by Sidgwick, who was disappointed that he didn't get a first-class degree in the new Moral Sciences Tripos, and having got to know one of Darwin's children and visited Down House (the Darwin family home in Kent), he was at rather a loose end on graduation. His uncle found him a safe seat in Parliament, and he wrote a

[11] Victor Shea and William Whitla, eds., *Essays and Reviews: the 1860 Text and its Reading* (Charlottesville: University of Virginia Press, 2000).

[12] E. G. Sandford, ed., *Memoirs of Archbishop Temple* (London: Macmillan, 1906), II, 419.

book, *Philosophical Doubt,* published in 1879.[13] It was to have had 'scepticism' in the title, but because in the 1870s that had come to mean agnosticism, and his target was science rather then religion, he changed it to 'doubt'. The book reads well, but aroused little attention. Balfour admired science, admitting that he lacked the application necessary for the precision measurements for which Rayleigh was famous; his central point however was that science (like everything else) is founded on faith. Tyndall in Belfast believed that scientists had the big picture: atoms, thermodynamics and evolution were the key to understanding the world, and the task now was to fill in the details – as Rayleigh was doing with his quantitative experiments. For Balfour, these were provisional generalisations that were indeed working wonderfully well, but had no firm basis:

> Science is a system of belief which, for anything we can allege to the contrary, is wholly without proof. The inferences by which it is arrived at are erroneous; the premises on which it rests are unproved. It only remains to show that, considered as a general system of belief, it is incoherent.[14]

He was particularly opposed to what was coming to be called 'scientific naturalism', a version of scientism in which all aspects of human life were in principle reducible to evolutionary factors. A lover (like Tyndall!) of beauty in all its forms, of music and literature, he found it utterly absurd to explain it away through natural selection. So, concluding that Naturalism, fast becoming a kind of dogmatic scholasticism, fails us, he went on (in good traditional conservative vein) to urge his readers not to fret about inconsistencies or incompleteness. Our minds are finite, and we should not expect certainties, or to understand how everything fits together.

In Salisbury's cabinet, Balfour became in 1887 Chief Secretary for Ireland, a crucial post as debate raged about Home Rule, amid Fenian outrages. To general surprise, the languid aesthete did very well in controlling explosive situations and pacifying the island; and in 1892 Balfour became First Lord of the Treasury and Salisbury's leader in the House of Commons. He was clearly the coming man, whose thoughts were worth reading. In 1894 he published *The Foundations of Belief,* again attacking Naturalism (in which he included Agnosticism, Positivism and Empiricism), a system in which we know phenomena and laws but no

---

[13] Arthur J. Balfour, *A Defence of Philosophic Doubt; being an Essay on the Foundations of Belief* (London: Macmillan, 1879).

[14] Ibid, 287

more: “All that gives dignity to life, all that gives value to effort, shrinks and fades under the pitiless glare of a system like this.”[15] His doubts now suddenly looked much more plausible, as previously unsuspected new rays and new elements (Rayleigh got a Nobel Prize for discovering argon) turned up in the 1890s. Cathode rays, the alpha, beta and gamma rays associated with radioactivity, and mounting difficulties over the nature of the æther, threw scientists into doubt. Balfour became Prime Minister in 1902, and in 1904 when the British Association for the Advancement of Science met in Cambridge, he was invited to be President. He drew attention to the way the physics that he had learned forty years before had been falsified. Though still unknown to those at the BAAS, Albert Einstein was demonstrating that Newtonian physics indeed rested upon doubtful beliefs about space and time, and that the received doctrine that energy was continuous was false. Moreover, physicists were quarrelling with geologists about the time available for evolution, and biologists were increasingly uncertain about how natural selection could explain all that was required of it. Balfour came to look like a prophet.

He fell from office in the Liberal landslide of 1906, but during the Great War as Foreign Secretary promulgated the declaration establishing a Jewish homeland in what is now Israel. In 1918, a delegation from the Royal Society invited him to be the next President; but finding that work was expected of him, he declined (and became President of the British Academy instead). The Society not merely looked for an elder statesman to steer them through the stormy waters of the post-war period, but also recognised his far-seeing scepticism. As things turned out, he was in effect the Minister for Science for much of the 1920s, as British governments continued the wartime support and promotion of scientific research and education. For a time, against the equally-aristocratic and well-connected Bertrand Russell, Balfour helped to maintain the position that science and religion were compatible, by his insistence that all knowledge rested upon faith, and must be incomplete, indeed incoherent.

As John Brooke’s paper shows, there are so many things going on in this fascinating period, when modern perplexities were first faced willy-nilly, that to find and hold on to threads that will guide us, and which twisted together will give a dense narrative, is difficult indeed.[16] There is a famous piece of advice to the historian: “read until you hear the voices”. Perhaps sometimes we nearly can – by the end of the period there are

---

[15] Arthur J. Balfour, *Foundations of Belief: being Notes Introductory to the Study of Theology*, 2nd ed. (London: Longmans, Green, 1895), 77.

[16] I tried to trace these through the long nineteenth century in David Knight, *Science and Spirituality: the Volatile Connection* (London: Routledge), 2004.

crackly recordings, but that's not what was meant. Out of the Babel we can at least hear the message that being a Christian Darwinian was possible, and for reasons for which it still is so.

PART II:

DARWINISM AND NATURAL THEOLOGY

CHAPTER SIX

NATURAL THEOLOGY AFTER DARWIN

DAVID FERGUSSON

The term 'natural theology' typically describes that branch of the subject that appeals to evidence for the existence and character of God from sources available to all rational beings. These usually include observation of the natural world as well as universal patterns of human experience – whether moral, aesthetic or religious. Excluded are appeals to particular sources of divine revelation, whether sacred texts, religious institutions, or single historical events. A methodological distinction is thus generated between a natural theology appealing to universally accessible phenomena and a revealed theology that deals with Scripture, tradition and miracles. From the early church onwards there is a tacit assumption that these two sources of knowledge, however configured, are complementary. Each supports the other, and together they offer a persuasive account of the being and action of God in relation to the creation. This yielded the famous metaphor of the two books – the book of nature and the book of God, their stories being combined to form a single, coherent narrative. In this way, it was generally assumed that scientific observation could support and confirm the fundamental convictions of faith, though we should beware of assuming anachronistically that these modes of reflection always represented clearly differentiated disciplines as in the modern academy.

Five types of natural theology

This stock description of natural theology will be familiar to most readers, and on the whole it is uncontroversial. Yet it conceals significant differences of approach and context. Natural theology has been used in a variety ways at different periods in the history of theology. By my reading of the subject, there are at least five types, and even some of these may be liable to sub-division. These range from stronger to weaker types of approach. What is interesting is that each has a particular historical context, conditions under which a strategy typically flourished as a response to various challenges and opportunities.[1]

1. First, we might cite the strong deist conviction that natural theology is a more reliable and less provincial guide to knowledge of God. A cause of much early modern conflict, claims based upon special revelation were perceived to be partial and divisive. Their epistemological status was less persuasive to an age increasingly impressed by the accessibility and universality of scientific claims. So we find a commitment to the priority of natural theology over revealed theology in writers such as Matthew Tindal at All Souls, Oxford. He argues for the superiority of natural theology. Scripture is to be assessed for its capacity to teach and conform to what we know on prior and independent grounds. At best, it contains a re-publication of the truths known to reason.[2] The existence of God, the providential ordering of the world and the moral truths which regulate our conduct can all be discerned by reason without reference to Scripture or the teaching of the Church. Indeed, much historical religion is here viewed as a corruption of an earlier and purer natural religion that was based upon reason and intuition. So the valorising of natural theology in deism is intended as a means of criticism of revealed religion. In this respect, it has a radical edge through much of the eighteenth century.

2. A second type of natural theology is less sceptical of revealed theology but it regards the former as necessary to establish the reliability of the latter. In other words, natural theology functions as a necessary preliminary activity upon which the claims of revelation can subsequently be founded. This role of natural theology is one that is sometimes found in modern epistemological projects. In a famous essay, Nicholas Wolterstorff

---

[1] I have borrowed this typology from an earlier essay, David Fergusson, "Types of Natural Theology," in *The Evolution of Rationality: Interdisciplinary Essays in Honor of J. Wentzel Van Huyssteen*, ed. F. LeRon Shults (Grand Rapids: Eerdmans, 2006), 380–393.

[2] "I desire no more than to be allow'd, that there's a Religion of Nature & Reason written in the heart of every one of us from the first Creation, by which all Mankind must judge of the truth of any instituted Religion whatever: and if it varies from the Religion of Nature and Reason in any one particular, nay, in the minutest circumstance, that alone is an argument which makes all things else that can be said for its support totally ineffectual." Matthew Tindal, *Christianity as Old as the Creation or the Gospel a Republication of the Religion of Nature* (London, 1731), 52.

detects this in Locke's claims for the reliability of Scripture.[3] These are based upon the manner in which belief claims generally must be justified. Hence a commitment to the truths contained in Scripture must be grounded upon the justification of the credibility of the Biblical writers. This is achieved, it seems, by reference to considerations about the veracity of testimony. Only as the witness of Scripture can be judged reasonable are we entitled to adhere to the distinctive truth claims of revealed theology. So the particular and essential commitments of Christian faith are grounded upon more general arguments for the divine existence and thereafter the reliability of Scriptural testimony confirmed by signs and wonders.[4] Once this is in place, we can then turn to the Bible with the knowledge that it will furnish us with further information about the works of God. Revelation can be believed as a source of knowledge not otherwise accessible, but only at the bar of natural reason. For Locke, this excluded the deplorable fanaticism and dangerous certitude of splinter groups.

3. A third type might be discerned in the work of Thomas Aquinas, although here interpretations vary. There are some truths about God which, although presented and known through Scripture, can also be attained by the use of human reason. Aspiring to such truth, the mind can comprehend in relative detachment from Scripture a knowledge of the existence and unity of God. This is a divinely-ordained goal of the human intellect even if not all human beings have the inclination, opportunity and capacity to attain it. For Aquinas, such knowledge does not precede Scripture in any kind of epistemological priority nor does it make Scripture redundant. In this respect, it performs a different function from the previous two types. Yet it appears to be a capacity of the human mind to reason from the world to God, at least in some rather limited ways. This function of natural theology can be found in the *Summa Contra Gentiles* alongside the apologetic task which is the next type. "Yet it is useful for

---

[3] Nicholas Wolterstorff, "The Migration of the Theistic Arguments: From Natural Theology to Evidentialist Apologetics," in *Rationality, Religious Belief and Moral Commitment*, ed. Robert Audi and William J. Wainwright (Ithaca: Cornell University Press, 1986), 38–81.

[4] The appeal to the miraculous events accompanying the writing of Scripture was long used as a supplementary mark of its divine authority. But for Locke it provides the rational warrant for believing Scripture to contain divine revelation. In this respect, revelation is epistemologically dependent upon the correct exercise of our natural reason. Cf. John Locke, *An Essay Concerning Human Understanding* (1690), Book IV, Chapter 19.

the human reason to exercise itself in such arguments, however weak they may be, provided only that there be present no presumption to comprehend or demonstrate. For to be able to see something of the loftiest realities, however thin and weak the sight may be, is … a cause of the greatest joy."[5]

Nevertheless, much recent interpretation of Aquinas understands the role of philosophical reason here neither as preparatory nor semi-independent, but as set firmly within the context of faith. The limits of reason are constantly stressed by Aquinas, while even the much-discussed Five Ways may be regarded as deflationary arguments to curb human pretension or idolatry. God's existence is not self-evident, and even the hard-earned conclusions of rational argument do not take us very far.[6] In this setting, natural theology has a legitimate, but clearly subordinate and circumscribed role. It is not an independent, autonomous source and norm for the knowledge of God. In this respect, Aquinas is not a forerunner of Locke or the deists.

4. A further function of natural theology is the more apologetic task of defeating the strongest objections levelled against some or all aspects of the Christian faith. In this respect, natural theology might include arguments against the eternity of the world, socio-biological explanations of morality, against the claim that religion is a form of false consciousness under the oppressive conditions of capitalism, and against the thesis that the freedom of the will is an incoherent notion. Although one can find arguments of natural theology in the early and medieval church performing this role, it becomes more prominent under the conditions of modernity where belief in God is no longer a given. However, its function here is mainly negative. It is designed primarily to "defeat the defeaters" to use Alvin Plantinga's expression. Attempts to convict believers of inconsistency, irrationality or sheer wishful thinking need to be confronted and rebutted. In this project, natural theology may have a role but it is clearly a defensive and subsidiary one, rather than a foundational or preparatory one.

---

[5] Thomas Aquinas, *Summa Contra Gentiles*, Book One, 8.1.

[6] "The point of insisting that arguments for God's existence is required is, then, not to convince hypothetical open-minded atheists, or even to persuade 'fools', so much as to deepen and enhance the mystery of the hidden God. From the start, the 'theistic proofs' are the first lesson in Thomas's negative theology. Far from being an exercise in rationalistic apologetics, the purpose of arguing for God's existence is to protect God's transcendence." Fergus Kerr, *After Aquinas: Versions of Thomism* (Oxford: Blackwell, 2002), 58.

5. Finally, a fifth task of natural theology might be discerned in the perceived need to display the ways in which the essential claims of revelation can co-exist in positive relation to the best insights available from other disciplines and fields of knowledge. Although closely related to the apologetic function, it has a more positive role in showing the consistency of theology with other important forms of enquiry, e.g. natural science and history. In this respect, natural theology remains a modest, ad hoc type of enterprise. It seeks to show the capacity of faith to co-exist with what we know of the world, its history and forms of life from other fields of enquiry not directly related to Scripture or the Church. In this last respect, natural theology may actually be viewed as an important pastoral and educational function of the Church. Active in many important secular activities – science, medicine and technology – the members of the Church seek to relate the knowledge and insights of the workplace to those of the faith communities to which they belong.

Given these different approaches to the function of natural theology, it is hardly surprising that its content should vary considerably also. What passes for 'natural theology' may include a wide range of claims, some more modest in scope than others. Deistic and Lockean arguments will typically make strong claims about the divine existence and nature which either precede or exercise their validity independently of Scripture, faith and the Church. A natural theology that reflects the divinely-bestowed powers of reason will tend to be determined by estimates of the strength and virtues of the human mind outside the visible Church. On one recent reading of Aquinas, noted above, natural theology may actually attain relatively little in this respect. Similarly, while Calvin attached some significance to the *sensus divinitatis* and classical arguments for God's existence, these were of minor significance compared to the theological tasks with which he was confronted. Clouded by the insidious effects of sin, our reason and judgement are unreliable without the guidance of Scripture and mother Church. When we come to the more defensive role of natural theology (type 4), we find that it is largely set by secular forces, these being combated by negative arguments to defeat the defeaters. Hume, Darwin, Marx and Freud presented challenges to theology that have been met in different ways but all of which were largely reactive. Yet, whether the outcome of this exercise leaves faith exactly where it was is doubtful. Its historical mobility and capacity to adapt under shifting intellectual conditions are not entirely captured by the language of defeat. Moreover, the ineluctable vocation of attesting that Christian faith can co-exist with the best insights from other fields of knowledge may largely be an ad hoc task, driven by the social and institutional context of the theologian.[7] In short, what we find is that the term 'natural theology' has been stretched to comprehend a wide variety of theological tasks that cannot be reduced merely to a reformulation of the traditional philosophical arguments for God's existence. It is regrettable that the way in which the proofs are frequently taught in school and university syllabi tends to obscure these important shifts in context and function.

Natural theology after *The Origin of Species*

Published in 1859, *The Origin of Species* is often represented as a watershed in the history of relations between science and religion. With the explanatory mechanism of natural selection, it was argued, much that had previously been attributed to the operation of divine design could now be explained by purely natural causes. Almost overnight, God had become redundant. Although not put about by Darwin himself, this view was encouraged by some of his followers, most notably T.H. Huxley, famously dubbed 'Darwin's bulldog'. For Huxley, the teleology favoured by Archdeacon Paley earlier in the nineteenth century could now be given natural explanation. The adaptation of means to ends throughout creation no longer required to be attributed to design. Organs had evolved with greater complexity, species had adapted to the environment, and changes had taken place in life forms under the pressure of natural selection rather than divine design. According to Huxley, what theology called providence, science perceived only as natural order.[8] This fundamental opposition was affirmed also by voices on the theological side.[9]

We find echoes of this dialectical opposition of scientific and religious explanation in much of today's new atheism. It is admirably captured in Daniel Dennett's metaphor of cranes and 'skyhooks'.[10] As the power of nature's cranes becomes better understood with the march of science, so the need to appeal to a skyhook recedes. Here the assumption is that science and religion occupy common ground in their explanatory accounts of nature. At earlier stages in the history of ideas, large and incomprehensible

---

[7] The term is borrowed from Hans Frei's justification of 'ad hoc' apologetics and correlationism. Cf. *Types of Modern Theology* (New Haven: Yale University Press, 1992), 70–91.

[8] Cited by David N. Livingstone, *Darwin's Forgotten Defenders: The Encounter between Evangelical Theology and Evolutionary Thought* (Grand Rapids: Eerdmans, 1987), 49.

[9] E.g. Charles Hodge, *What is Darwinism?* (London: Nelson, 1874).

[10] Daniel Dennett, *Darwin's Dangerous Idea* (New York: Touchstone, 1995), 136.

gaps in our understanding of nature were filled by recourse to a divine skyhook. However, as these gaps diminish so the cranes start to do the heavy lifting, with the result that religion becomes ever further marginalised. Richard Dawkins is the most strident exponent of this view today. Once upon a time natural theology had sought to explain the appearance of complex organs such as the eye, as well as the extraordinary adaptation of species to environment. But now a neo-Darwinian account of genetic mutation and natural selection can offer a powerful story of how this all came to be. Darwin's brilliant idea works like a powerful acid upon earlier forms of theological reflection. Instead of a worldview dominated by the hypothesis of intelligent design, we now have a naturalist outlook in which God lacks any substantial explanatory role. Nevertheless, despite its powerful advocates, the view that Darwinism inevitably leads to the dissolution of theology is neither historically nor philosophically compelling. Soon after the appearance of *The Origin of Species*, Christian theologians showed themselves both persuaded by the force of Darwin's science and ready to offer it theological accommodation. Many of their arguments adumbrate positions adopted by scientists and theologians today in relation to evolutionary theory.

Given the ways in which Darwin's theory of evolution built upon a significant body of scientific discovery in the first half of the nineteenth century, we should not be surprised to discover that theologians were already somewhat prepared for the subsequent debate. Charles Lyell's work in geology had persuaded many that the Earth must be millions of years older than the traditional dating of 4004 BCE proposed by Archbishop Ussher on the basis of the Biblical record. By the early nineteenth century, geologists could show that through 'uniformitarian' processes the formation of rocks and changes to the Earth's surface could be explained by natural causes such as volcanoes, glaciers, sedimentation and erosion by water. These processes, however, required a great deal of time and only a much older universe than the one posited by Ussher would fit the bill. From about 1820, therefore, we find many theological writers denying a young universe that had been decisively shaped by a catastrophic event, i.e. the flood. The capacity to interpret Genesis Chapters 1 to 11 in non-literal ways was already apparent, as of course it had been in the exegesis of the early Church Fathers.[11] As always, the intellectual causes for a sudden shift in perspective are complex and multiple. Why then did Darwin's theory seem to cause such an intellectual

---

[11] Symbolic readings of the Genesis story were commonplace in the early church in any case, as is evident for example in Augustine's Confessions.

upheaval for religion? At least, four reasons are apparent, each will be explored.[12]

i) *God now appeared to be remote from the cosmos.* The natural theology that had dominated British theology until the early nineteenth century was epitomised by William Paley. Here the evidence of divine design was everywhere manifest, so much so that little credence was given to Hume's scepticism advanced earlier. While the arguments of Philo in Hume's *Dialogues* may have appeared unassailable to later analytic philosophers, for much of the nineteenth century they were regarded as overblown and fanciful. The evidence of design was too compelling to sustain such scepticism. Successive contributions to the Bridgewater Treatises, including that of the evangelical theologian Thomas Chalmers, confirmed this confidence in the design argument.[13] So much of the natural world was inexplicable without appeal to the God-hypothesis that it was assumed that a greater understanding of its workings would only confirm the intuitive impression of design. But if natural selection could explain what had formerly been assigned to the mechanism of supernatural design then the latter would become redundant. This was undoubtedly a fear in some theological quarters matched by the hopes of Huxley et al. Yet, this placing of God on the unemployment register was much too hasty. Since at least the time of Newton, scientists and theologians, impressed by the explanatory power of scientific law, had seen its rational operation as evidence of the majesty of the Creator. Already we see here a conviction about the consistency of belief in a divine Creator with explanation according to scientific law. If the laws of nature could explain the formation of rocks, then this merely registered the power of the Creator who had ordained those laws.[14] One way of understanding this theological appropriation of natural science is to note that what was formerly attributed to the working of special providence was now assigned to a general providence. God was not required occasionally or frequently to intervene in the cosmic process in order to achieve the intended results. Instead, the manner in which the world had been established under the

---

[12] In what follows, I have drawn from an essay on "Providence after Darwin'", in *Theology After Darwin*, ed. Michael Northcott and R. J. Berry (Milton Keynes: Paternoster: 2009), 73–88.

[13] Thomas Chalmers, *On the power, wisdom, and goodness of God as manifested in the Adaptation of External Nature to the Moral and Intellectual Constitution of Man*, 2 vols. (London, 1833).

[14] See the argument of John Hedley Brooke, "Natural Law in the Natural Sciences: the Origins of Modern Atheism?," *Science and Christian Belief* 4 (1992): 83–103.

general working of natural law was itself sufficient to realise those creaturely states and entities desired by God. Not surprisingly, this move was to become a stock response to Darwinian theory. If evolution is how states of greater complexity emerge in the history of the cosmos, then it is open to the theologian to claim that this is how God does it. In his 1884 Bampton Lectures, we find Frederick Temple adopting this strategy. He writes that God "did not make the things, we may say: no, but He made them make themselves."[15] This was teleology at one distance removed and it offered a different account of design to that of Paley. Indeed in an important respect, Temple sees his view as improving upon Paley. Instead of an Artificer who has to interject at regular intervals to bring about the intended effects in the production of life forms, we have a Creator who has from the very beginning endowed the creation with sufficient natural powers to evolve as intended.

This argument was already employed by Principal Robert Rainy in his inaugural address at New College, Edinburgh in 1874.[16] According to Rainy, the patterns of an evolving world require to be explained by reference to divine design. All that has changed is our conception of a world that is originally endowed with sufficient fruitfulness to yield these emergent patterns. Rainy notes that there may be some loss "of the argumentative benefit of pleading earlier interpositions as analogical instances" of divine revelation in history. A God who regularly intervenes to direct the course of history might be expected to do the same in the natural world. However, this assumption can be yielded in favour of an evolutionary world-view. While expressing some reservations about Darwinism, he seeks to distinguish the approach of the natural scientist from that of the theologian. These different forms of understanding occupy separate domains, allowing a relative independence within each but a complementarity when viewed in conjunction. One upshot of all this is the clearer recognition of the disciplinary boundaries between theology, philosophy and the natural sciences.

It is worth noting that already embedded in this approach is the best response today to Creation Science and Intelligent Design theory. An evolving world, as described by the natural sciences, can equally well be viewed as the outcome of divine design. We do not need to seek gaps in scientific explanation in which to interject the agency of God. By allowing these spaces to be filled by subsequent scientific research, we can concede this domain of explanation to the natural sciences. At the same time, the

[15] Frederick Temple, *Religion and Science* (London: Macmillan, 1884), 115.
[16] Robert Rainy, *Evolution and Theology* (Edinburgh: 1874).

theologian can insist that God has endowed the creation with the capacity to evolve increasingly complex patterns of life forms.[17] To appeal to divine intrusion at the point when the latest scientific explanation falters is to give hostages to fortune. As science progresses so the gap is closed. In any case, such defensive strategies fail to recognise the different and complementary levels at which physical and theological explanation operate.[18]

ii) *The role of chance and loss of providential control.* Much theological anxiety can be detected around the role assigned to 'chance' by Darwinian science. Even amongst writers in search of an alliance with evolutionary theory, we find attempts to replace the function of chance with a more deterministic mechanism. Of course, for Darwinian theory 'chance' does not refer to the inexplicable or the uncaused. Instead, it is the denial of a single deterministic trajectory followed by the evolution of life forms. There seem to be two types of process that are characterised by the language of chance. One of these concerns the minor physiological variations that are evident through the reproduction of species, and the other is in the intersection of unrelated causal systems, e.g. the impact of a sudden change in climate upon the development of species in a hitherto stable ecological niche. The mechanism governing physiological variation was not understood until the later development of genetics, and many informed critics of Darwin in the nineteenth century recognised this lacuna in his theory. It was eventually to be filled by developments in genetics, thus providing the neo-Darwinism synthesis with its twin principles of natural selection and genetic mutation.

Yet it was the random course of evolution, as described by Darwin, that most offended Charles Hodge in Princeton. He regarded this as practically atheistic since there could be no governing purpose or overriding control exercised over the direction of nature. What was under threat here was not so much the doctrine of creation from Darwinian evolution – a transcendent origination of the whole scheme could still be conceived – as the doctrine of providence. Hodge could concede that a process of evolution was consistent with theism. However, the particular account offered by Darwin with its stress on natural selection led him to believe that it was metaphysically inconsistent with the teleological principle that belonged both to revealed and natural theology. If God were

[17] This is the response favoured by Howard van Till, "The Creation: Intelligently Designed or Optimally Equipped?," *Theology Today* 55 (1998/9): 344–364.

[18] This is eloquently argued by Francis Collins, *The Language of God: A Scientist Presents Evidence for Belief* (New York: Free Press, 2006).

no longer in control of the course of life on earth, then it could not be perceived as proceeding towards an appointed end. For a Reformed theologian such as Hodge, this was tantamount to a practical atheism.

Nevertheless, within that same tradition others reached different conclusions by placing evolution within a wider context that was perceived to be providentially ordered. Asa Gray, the Harvard botanist, replied to Hodge. His difficulty with Darwin, Gray claimed, arose out of an unduly restricted account of how divine teleology works. The Creator can endow nature and organisms with the powers of evolution into states of greater complexity. Hodge simply begs the question against this type of teleology with its commitment to an older, Paleyian type of strategy. This was also the line taken by Robert Flint in his Baird lectures on *Theism* that went through numerous editions during the late nineteenth century. Here his interaction with Darwinism offers a set of responses that have become standard in theological appropriations of evolutionary science. The development from lower to higher organisms can be explained as a mark of design. The tendency towards improvement and progression requires explanation. Evolutionary process can be envisioned as a vast scheme of order and beauty, rather than a grim arena of conflict and waste. This more positive vision of evolutionary complexity would be developed by writers such as J. Arthur Thomson and Patrick Geddes in their 1912 study of evolution.[19]

Later in Princeton, Alexander Hodge and B.B. Warfield would make their peace with Darwinism, arguing that it was scientifically sound but also consistent with an account of divine providence. What much of this shows is that there was no clear consensus about Darwinism at that time. It was interpreted in a wide variety of ways so that even to talk of Darwinism, as if it meant one thing to everyone, is quite misleading. James Moore, biographer of Darwin, has argued that in the 1860s there were as many as five different types of Darwinism, each contending with one other.[20]

Other theological responses after the publication of *Origin* offered a more immanent account of divine involvement in the evolutionary process. This suggests a different model of providence and again it is one that continues to attract defenders today. In his essay in *Lux Mundi* (1889), Aubrey Moore wrote in an oft-quoted passage:

---

[19] J. Arthur Thomson and Patrick Geddes, *Evolution* (London: Williams & Norgate, 1912).

[20] James Moore, "Deconstructing Darwinism: The Politics of Evolution in the 1860s," *Journal of the History of Biology* 24 (1991): 353–408.

> The one absolutely impossible conception of God, in the present day, is that which represents him as an occasional visitor. Science has pushed the deist's God further and further away, and at that moment when it seemed as if He would be thrust out all together Darwinism appeared, and, under the disguise of a foe, did the work of a friend.[21]

Moore's claim is that Darwinism affords a positive opportunity for Christian theology to re-affirm its commitment to a perpetual involvement of the divine Spirit in the natural world. Instead of the more remote and disengaged God of deism, we now have an immanent divine presence that is at work in the creative and open-ended processes of evolutionary history. Rather than posing a threat, therefore, the apparently free movement of natural forms is consonant with a God who is present and active within the creative process, in a manner analogous to God's same involvement with human history. This model of divine engagement resonated with the kenotic theologies that flourished in the late-nineteenth century and it has exercised a particularly strong hold over Anglican thinkers to the present day with their stress on divine passibility. The model here is of God's "letting the world become itself," not in such a way as to abandon it but in the interests of a patient accompanying that seeks to work within and alongside creative processes. A model of providence is thus suggested that avoids the perceived determinism of much of the Augustinian tradition but also the deism that had persisted from the early modern period. What is interesting about this reading of evolution, for the purpose of this essay, is the way in which science is used here not so much to demonstrate theological conclusions but rather to show the consonance or fit between evolutionary theology and a kenotic doctrine of God. A more modest approach of discerning a possible harmony of understandings is now typically pursued.

More than most who sought an accommodation with Darwinism, Aubrey Moore was able to see clearly some positive theological gains. The occasional interventions of a remote deity could now be replaced by the constant sustaining and creative activity of the divine Spirit. This has the significant benefit of correlating the Christian understanding of how God acts in nature and history. The God of the Bible is portrayed as deeply involved in the stories of Israel, Christ and the Church, an ongoing personal drama that takes a narrative shape in the successive books of Scripture. By contrast, the model of divine action suggested by Deism, at least in one of its standard forms, was that of a transcendent Creator who

---

[21] Aubrey Moore, "The Christian Doctrine of God," in *Lux Mundi*, ed. Charles Gore (London, 1889), 99.

perfectly ordered the world so that subsequent historical interaction was unnecessary, even undesirable. Given the way in which this mode of divine action contrasted with the standard scriptural account, it was hardly surprising that deist writers were more or less sceptical of historical revelation. At best, revelation provided a re-publication of the truths of natural religion.[22] While Paley's teleology did not amount to a wholesale commitment to Deism, its model of the God-world relationship seemed to lean in that direction with the notion of a perfectly ordered world in little need of repair or further development. After Darwin, however, theologians could see the world of nature as having a history, as being constantly in the process of making. It had a narrative shape that could increasingly be detected by advances in the natural sciences. Like history, therefore, nature was a work in progress, a construction site on which God could be seen as a sustaining, creative, guiding presence. This appearance of the natural world as undergoing significant change across time was of course further confirmed by the emergence of big-bang cosmology in the following century.

iii) *The intensification of the problem of evil.* A further recurrent concern in the theological reception of Darwin concerns the problem of evil. Of course, this was hardly a new challenge to confront Christian theology. The Book of Job already reveals a long history of reflection on this in Jewish traditions. And, in any case, the facts of disease, suffering and death were manifestly visible before any theory could explain their contribution to the evolutionary story. But what was striking in Darwin's description of evolution was the extent to which suffering, waste and the competition for survival were the drivers of evolution. These were part of the 'design' that enabled the emergence of species, including human beings. Instead of Paley's notion of creatures living always in a state of equilibrium, their prosperity secured by a single divine blueprint, theologians now faced a bleaker scenario in which earlier species were driven to extinction in a perpetual warfare of life forms. How could this suggest a divine providence?

One type of theodicy that quite confidently took Darwinian explanation on board argued, more or less, that the end justified the means. Since evolution produced fitter and more advanced species, particularly *Homo sapiens*, we could conclude that the laws of evolution were all part of a benevolent divine plan. The production of better adapted forms of life could thus be seen as an outworking of an overall teleology. This was the

---

[22] See for example Tindal, *Christianity as Old as the Creation.*

strategy pursued by the aforementioned Duke of Argyll and Henry Drummond in Scotland, and also Temple in his Bampton Lectures.[23] At the same time, these writers sought to introduce a different set of metaphors to overcome the prevailing sense of waste and random, meaningless suffering. Attention is drawn to the interdependence of species, the unity of the natural world, the long periods of relative equilibrium characterised in large measure by the enjoyment of life, and the beautiful harmony of flowers and insects.[24]

Others would point to patterns of self-sacrifice in nature that adumbrated later Christian moral ideals. In his study *Christianity and Evolution*, James Iverach noted that the individual often sacrifices itself for the wellbeing of the species to the extent that something like a virtuous family life emerges naturally.[25] One danger inherent in this kind of teleology is that works with a principle of the end justifies the means. If nature discards the weak and the unsuccessful, then this is so that fitter and better-adapted species may emerge. The ends of evolution enable us to explain why the system is one in which much is discarded as unfit, superfluous, and disposable. This could tilt in dangerous directions particularly as human evolution came into focus. Should social policies be developed that artificially selected the fittest for reproduction? We are not so very far here from theories of social Darwinism and the eugenic theories and practices that were evident in the early twentieth century, and not just in Nazi Germany.[26]

Nevertheless, by the later nineteenth century, this rather confident style of theodicy was being replaced by more reserved approaches to the problem of suffering. While a divine plan may be in a process of outworking, its meaning and significance are not yet discernible from a human vantage point. We simply cannot speculate on what purpose long aeons of animal suffering may serve. Again, however, a virtue can be made out of a necessity. The theologian has no business reading off the details of divine design from the pages of natural history. The only index to providence is that of faith in Christ – more speculative and comprehensive accounts should be eschewed. At the same time, Darwinism

---

[23] See the discussion in James C. Livingston, *Religious Thought in the Victorian Age* (London: T & T Clark International, 2006), 69ff.

[24] Temple, *Religion and Science*, 122.

[25] E.g. James Iverach, *Evolution and Christianity* (London: Hodder & Stoughton, 1894).

[26] For further discussion see Diane B. Paul, "Darwinism, Social Darwinism and Eugenics," in *The Cambridge Companion to Darwin*, eds. Jonathan Hodge and Gregory Radick (Cambridge: Cambridge University Press, 2003), 214–239.

may also helpfully save the theologian from embracing too narrow an anthropocentrism. Given the relatively late emergence of human beings and the extent to which animal life has evolved for much of the time with no reference to ourselves, we cannot assume that God's purposes are solely directed towards our own species. God must have more in mind than the evolution of humankind.[27]

iv) *The perceived threat to human significance.* This last point brings us to another hotly-debated topic emerging from Darwinism, namely the significance of human beings in creation.[28] To a large extent, the fear surrounding evolutionary theory was that it threatened to undermine our most fundamental convictions about the intellectual, moral and spiritual distinctiveness of human being. This was expressed immediately after the publication of *Origin of Species* in 1859 and before the appearance of the *Descent of Man* in 1871. Indeed, this fear continues to infect much of the debate today. If human beings emerged from other mammals over millions of years of evolution, is their uniqueness thereby destroyed? Again a religious accommodation with Darwinism was proposed by successive writers, while also maintaining a traditional commitment to human distinctiveness. Scholars like Rainy and Temple argued again for a complementarity of explanations. It was open to the scientist to point to the ways in which human beings had emerged from other primates while also displaying significant similarities to other species. Yet the theologian could also point to phenomena such as consciousness, our moral capacities, and spiritual discernment that demand different forms of explanation and understanding from those available to the natural sciences. Here natural theology in the late nineteenth and early twentieth centuries leaned more in a Kantian direction with its preoccupation with moral phenomena. We see this for example in the arguments for a moral theism in many of the early Gifford lecturers.

Natural theology in the twentieth century

On the basis of this nineteenth century survey, some concluding remarks may be ventured on the fate of natural theology through the twentieth century. It is clear that Darwinian evolution did not sound the death knell for the subject. Around the time of Darwin's death, Lord

---

[27] Aubrey Moore, "The Christian Doctrine of God," 108f.

[28] For recent discussion see J. Wenztel van Huyssteen, *Alone in the World? Human Uniqueness in Science and Theology* (Grand Rapids: Eerdmans, 2006).

Gifford was preparing his last will and testament that would ensure that natural theology was studied and defended before a public audience in each of the four ancient universities of Scotland. The history of the Giffords points to the vitality of the subject and its capacity to flourish under shifting intellectual conditions. While Darwinian science rendered one version of the design argument redundant, physicists soon became fascinated with instances of so-called fine-tuning which suggested that ours is a universe with an anthropic bias. The debate continues, now with intense discussion of the multiverse concept.

The revival of the philosophy of religion in analytic circles, championed by scholars such as Swinburne, Stump and Plantinga in their different ways, has occasioned greater confidence in some of the standard arguments of natural theology. Apologists such as William Lane Craig in the USA now argue confidently (too confidently in my view) that big bang cosmology confirms the ancient kalam cosmological argument that the universe must have had a temporal beginning, this being the consequence of a divine creative act. At the same time, the emergence of comparative theology has drawn attention to the ways in which scholars in each of the Abrahamic faiths have tended to share similar arguments and strategies in defending the convictions of their communities.

Much of this revitalised natural theology, however, has viewed science and theology as complementary types of understanding, not competing with each other to explain the phenomena but respecting the particular questions, domains and procedures of these different forms of enquiry. This clearer differentiation of disciplines was in part an outcome of the engagement with Darwinism; increasingly it was recognised by theologians that it was futile to attempt to contest scientific explanation in order to save the day for God.

Finally, what of the total resistance to natural theology on the part of Karl Barth and many of his acolytes? We should note here that Barth's rejection of the discipline continues a long history of Germanic scepticism towards natural theology that goes back to Kant and Schleiermacher, who regarded the standard proofs as both unconvincing and anthropomorphic. This hostility was shared by the liberal school of Albrecht Ritschl which viewed *history,* rather than *nature,* as the locus of divine activity. Metaphysical speculation was eschewed in favour of historical study and the interpretation of the distinctively ethical dimension of human existence. Emerging from out of this tradition but turning against it, Barth launched a positive theological attack on natural theology. It was not just hopeless but actually sinful in principle. The proofs were a decidedly bourgeois exercise that could yield only a God in our own image, belief in whom was a form of self-aggrandisement and acquiescence in the status quo. Much of this, especially his famous 'Nein' to Brunner, was played out in the heat of the German Church controversy in the early 1930s. Natural theology is even denounced as the anti-Christ, leading us astray from proper obedience to the one Word of God who is Jesus. The only God who is self-disclosed in a determining act of revelation is the God and Father of our Lord Jesus Christ.

Barth's theology clearly excludes some types of natural theology, particularly those that assign to it a foundational, indispensable or controlling function. But whether it completely forbids the more modest strategies of defeating the defeaters or showing the compossibility of theological with other forms of knowledge remains an open question. At any rate, in the work of T.F. Torrance, his leading British pupil and commentator, we find an attempt to revive natural theology by integrating and subordinating it to the claims of positive theology. And in Barth's own later writings, his discussion of the little lights of creation may offer possibilities of more nuanced interpretation of his prohibition of natural theology.

Where does this leave us? It is clear that increased awareness of the contextual setting of all natural theology, together with our greater sense of the disciplinary discreteness of science and religion, must render natural theology a more modest type of enquiry than that supposed by either the Deists or Archdeacon Paley. Natural theology takes place within a setting of presuppositions, particular problems and challenges that require to be addressed. Its shifting contexts suggest that it has a more *ad hoc* character in intellectual debates than was often recognised. At the same time, a recognition of science and religion as offering complementary forms of understanding should caution against any assumption that there is a direct argumentative line from scientific premises to theological conclusions. This too will tend to ensure that natural theology assumes a more modest and deflated character, despite some contemporary attempts to move swiftly from big bang cosmology to creation out of nothing. In any case, we need to recognise that faith is not so much the result of a particular propositional state generated by an argument. It is a wide-ranging commitment of a practical, affective and intellectual nature entered into as a result of diverse considerations not readily capable of precise theoretical expression.

The issue of the relationship of faith to reason greatly exercised John Henry Newman in his *University Sermons* and later in the *Grammar of Assent* around the time of Darwin's scientific discoveries. Pointing to the practical character of religious belief, he claims that faith is mainly swayed by 'antecedent considerations'. These are its "previous notices, prepossessions, and (in a good sense of the word) prejudices."[29] Preaching in 1839 on the text that God has chosen the foolish things of the world to confound the wise, Newman compares the judgements of faith to the wisdom of a great general who "knows what his friends and enemies are about, and what will be the final result, and where, of their combined movements." Yet when asked to argue in word or on paper, the same general may find all his conjectures and reasonings to be less than adequate. By analogy, he speaks of:

> faith (as) a process of reason, in which so much of the grounds of inference cannot be exhibited, so much lies in the character of the mind itself, in its general view of things, its estimate of the probable and the improbable, its impressions concerning God's will, and its anticipations derived from its own inbred wishes, that it will ever seem to the world irrational and despicable – till, that is, the event confirms it.[30]

Natural theology may have a place in this scheme of things, but it will tend towards a subordinate role in relation to the wider judgement of faiths, and may be none the worse for that. So in terms of my initial typology, the first two types may now be excluded whereas the latter two exercise, *pace* Karl Barth, a significant auxiliary role in defending and declaring the faith. The third type too may have its place – people of who sit quite lightly to any faith community or tradition will continue to inspect and sometimes to rehearse the arguments of natural theology. In our own time, scientists such as Paul Davies have done this to good effect.

It seems then that neither Darwin nor Barth in their different ways succeeded in killing off natural theology; perhaps it was only particular historical manifestations that aroused scepticism in the former and anger in the latter. Given the variety of types of natural theology and its adaptation to different historical contexts, its capacity readily to mutate should not surprise us.

[29] John Henry Newman, *University Sermons* (London: SPCK, 1970), 187.
[30] Newman, *University Sermons*, 217–8.

CHAPTER SEVEN

RESPONSE: NATURAL THEOLOGY AND THE FLAT-PLANE FALLACY

SARAH COAKLEY

First, many thanks to you, David, for a clear and thought-provoking set of distinctions on five types of 'natural theology', and an – at least implicit – tipping of your hat towards types 4 and 5, *pace* Barth.

I would like to make just three broad comments in response to your lecture, in order to open the discussion:

1. You indicate at one point that these 'types' could do with further subdivisions, as 'types' always do. In this regard I would be particularly keen to hear your views about the further subdivisions as you see them within type 3, which involves a reading of Thomas Aquinas now popular and influential, to be sure, but not a reading with which I myself wholly concur. (One is bound to ask: "If Thomas meant what Fergus Kerr thinks he meant – viz., that the arguments for the existence of God are simply there to instil a proper sense of 'mystery' – then why, in the *Summa Theologiae* – does he not just say that?") In these post-de-Lubacian days, not to say post-*Radical Orthodoxy* days, we have come to be coy about enunciating the distinctions between philosophy and *sacra doctrina* which Thomas does himself clearly make in the opening sections of the *Summa*. I would be glad to hear what Professor Fergusson really thinks about this, because I think he stepped around the hermeneutical issue rather neatly in what he gave us tonight. But quite a lot hangs on it, as may emerge in my other two comments.

2. We detect that Professor Fergusson feels ambivalence about type 3 'natural theology', but some guarded enthusiasm about types 4 and 5. If the enthusiasm *is* guarded, it is only because the shadow of Karl Barth still

hangs over the proceedings; and Scottish *pietas* seemingly dictates that Professor Fergusson cannot simply see off Barth's wholesale rejections of 'natural theology' without at least a polite genuflection! (However, I'd like in antithesis to press an immediate point here: if one of the good effects of this lecture is to make us more aware of the particular historical circumstances in which 'natural theology', or rejections of it, were *originally* enunciated, then we should stress the same point in relation to Barth's critique. If he were alive today, and setting out to counter the brash confidences of the New Atheists, I do wonder if his strategies would have been different.)

Be that as it may, my second major point is this. Your typology makes me newly anxious about the current state Christian philosophers and theologians are in when they come into conflict with secular philosophies of science. For what your typology underscores afresh is the degree to which we Christians are currently in theoretical *retreat* – as was charged memorably by Dan Dennett in an intervention at the Cambridge Darwin festival a few weeks ago. If our argumentative eggs are all in the baskets of type 4 and 5 'natural theology', they are – in your own words – 'weak' and 'modest' in status. Merely to do a defensive reaction in response to major reductive moves, on the one hand, or to declare Christian faith 'compossible' with scientific endeavours, on the other, are – whilst certainly not time-wasting for the believer – hardly strategies likely to convince the sceptic, let alone the atheist. At very best they establish that I, as a Christian, have not actually flouted any epistemic duties by continuing to hold onto some ostensibly comforting, but ultimately un-argued, Scriptural mythologies. You can see, therefore, why the remaining – and yet unclarified – status of type 3 becomes yet more significant to me in this light. And that is why I press you on it.

Let me put this second query in a different way, and thus: What we probably all can now agree on is that the strategies of types 1 and 2 'natural theology' made a fundamental mistake about a 'flat plane', a universal one, which we could supposedly all *meet* upon – scientists, philosophers, theologians – to hash out the truth according to some agreed epistemological rules. Let us call this the 'flat plane' fallacy. But what worries me about your alternative, Professor Fergusson, is that it seems implicitly to shore up that fallacy once more in effectively conceding the flat plane to *secular* philosophy and science and then just working around it with some minor theological extras. That is: the supposed flat plane goes on, but theologians and post-modern philosophers of religion of types 4 and 5 now dance around it, either declaring that it does not actually quite succeed in *some* of its more daring reductionist moves, or that we would like to add a different sort of interpretation that is at least 'compossible' with the flat planer's vision (e.g., that my belief is simply a personal preference, a sort of neo-Kantian overlay of theistic *interpretation*; or – alternatively – that my God cannot be seen at work by the flat-planers in the world because he is a 'kenotic' God and voluntarily hiding himself).

Even type 3 exponents of natural theology of the sort that Professor Fergusson reminded us (interpreting Thomas à la Fergus Kerr, David Burrell, *et al.*) seal themselves off hermetically from direct confrontation on the flat plane by declaring that the divine mystery is operating on an altogether *different* plane from the scientific one. In this they are of course right, ontologically and theologically speaking. Yet their strategy – as I read them – deliberately avoids direct apologetic challenges to the flat plane *in its epistemological flatness*. The two 'magisteria' of theology and science (as Stephen Jay Gould described them) go on at their different ontological levels, so much so that as sophisticated a Catholic philosopher of science as Ernan McMullin can write in a recent essay that, "There is nothing about the evolutionary process in itself that would lead one to recognize in it the deliberate action of a … Planner."[1] In short, no news from the flat-planers could either *suggest* planning *or* undermine it. If that is where the 'mysterious' reading of Thomas goes, then it certainly lets one off the hook of any possibility for robust, critical, engagement with the sceptical flat-planers. Type 3, on this reading of Thomas, in fact really becomes a disguised version of type 5, a mere compossibility option.

So my second question presses: Is Dan Dennett right when he says that we Christians are for the most part – whether Barthians, de Lubacian Thomists, or liberal neo-Kantian Protestants – just in radical intellectual retreat on the science front? And if we are, are we not – apart from a few minor tweaks via type 4 approaches – pretty much leaving the flat plane of atheistical science where it is?

3. My third and final question follows immediately from this. It seems to me that a new version of type 3 needs to be devised that is much less coy about its challenge to atheistical science than your preferences, Professor Fergusson, seem to suggest. I will not go back here again to my first, hermeneutical point (how to read Thomas on the relation of philosophy and theology) although that is clearly related. It is however separable from the question of principle I want to raise here. What I would like to suggest is that we need a two-fold, pincer-like, move to challenge

[1] See Ernan McMullin, "Cosmic Purpose and the Contingency of Human Evolution," *Theology Today* 55 (1998): 389-414.

the complacency of atheistical neo-Darwinists, and to destabilize the flat plane that they still believe themselves to enjoy commanding. The first move, the ontological move, is indeed the classic Thomist one – as expressed by Kerr, McMullin and many others – viz., that whatever God's actions involve they are operative at a completely different ontological level than that of discrete created events. This position already complicates the flat-plane position by insisting that it is not the only, or finally undergirding, scene of operation; and the importance of this first point can scarcely overestimated. In many 'science'-religion debates it is still not adequately understood. God does not compete for space with individual events studied by the scientist.

However, the mere *claim* that this deeper ontology exists will scarcely in itself convince the sceptic if no further reasons can be given for it except that it invokes appropriate 'mystery'. So this ontological destabilization of the flat plane dogma needs to be complemented – in my view – by a different, and sophisticated, epistemological and hermeneutical one, which really takes on working scientists where they are, and, as a matter of apologetics, challenges them to explain the metaphysical and narrative presumptions on which they themselves operate. Much good preliminary work has of course been suggested along these lines, as Professor Fergusson notes, by the 'defeating the defeater' exponents. But I think this work could afford to become more daring hermeneutically, by probing the ways in which certain practices of devotion, on the one hand, or certain religious stocks of metaphors and narratives, on the other, actually can still activate and enable particular scientific developments of insight which – one could argue – would not be *likely* without them; and yet such insights can come to command full respect in the scientific community when they are worked out in standard 'scientific' terms. (I could give an example here from my work with the Harvard evolutionary biologist Martin A. Nowak. In his ground-breaking discoveries about the importance of the 'principle of cooperation' in evolutionary development, he has most certainly been animated and inspired in his research by certain stock Christian metaphors such as 'sacrifice', 'forgiveness', and so on.[2]) In other words, the case has to be made, using actual pressing working examples, of how the scientific imagination and processes of understanding have been, and in some important cases continue to be, implicitly activated and sustained by religious practices and insights which

[2] For a brief introduction to Nowak's theory about the significance of 'cooperation' alongside mutation and selection, see his "Five Rules for the Evolution of Cooperation," *Science* 314 (2006): 1560-63.

have given accesses to the real not otherwise easily glimpsed. Is this a tall order? Certainly. Is it a *too* tall order? I do not think so.

Note that this suggestion is not at all an invitation to return to the *old* 'flat plane' (whilst attempting to sneak theology into it). On the contrary, it is precisely complicating that picture of a flat plane by the insistence on distinct hermeneutical options in scientific endeavour, ones which can be rationally debated, to be sure, but not ones which can simply be reduced to the insistence on a mandatory methodological atheism. And if such an epistemological confrontation is somehow deemed theologically off limits, then – I ask again in closing – how can 'natural theology' now be anything more than a series of defensive, and quite optional, intellectual games, of which – rather obviously – the sceptic will feel s/he has 'no need' whatever? In other words, if all we have is types 4 and 5 of 'natural theology', what prospect for the intellectual and rational impact of theology in the university in coming decades? The matter seems to me to be one of intense urgency. It is not good enough, I submit, to embrace 'modest', optional, interpretative glosses of a theological kind on normatively atheistical science; there has to be a more convincing and creative alternative for Christianity today.

CHAPTER EIGHT

A CRITIQUE OF INTELLIGENT DESIGN

DENIS ALEXANDER

Intelligent Design (ID) has been critiqued many times before.[1] The purpose of this chapter will be to review a more recent publication from an ID proponent on cell biology to determine whether there have been any significant advances in ID thinking since the movement first came to prominence, and to assess the arguments made from scientific, philosophical and theological perspectives. The book to be scrutinized is *Signature in the Cell* by Stephen Meyer.[2]

A few initial comments on terminology may be useful. The term 'intelligent design', spelt all in lower case, is sometimes used to refer to the general argument from design familiar as part of natural theology in which the existence or even attributes of God are claimed to be inferred from particular properties of the natural world. For those speaking from this perspective, it would be unexpected to find any particular scientific theory in conflict with such 'intelligent design', given that it refers to the intelligibility of the universe and its properties that render the scientific enterprise feasible. By contrast, Intelligent Design refers to a specifically anti-Darwinian movement, distinct from creationism, that emerged in the USA during the early 1990s. The flagship centre of the movement is the Discovery Institute of the Center for the Renewal of Science and Culture located in Seattle, Washington State, USA. Apart from the conviction that Darwinian evolution represents a naturalistic philosophy, and therefore should be opposed, the other main theme uniting an otherwise somewhat

---

[1] Kenneth R. Miller, *Finding Darwin's God* (New York: HarperCollins, 2000); Francis S. Collins, *The Language of God* (New York: Free Press, 2006); Denis R. Alexander, *Creation or Evolution - Do We Have to Choose?* (Oxford: Monarch, 2008); Denis R. Alexander, "Is Intelligent Design Intelligent?" in *Theology After Darwin*, eds. R. J. Berry and M. Northcott (Milton Keynes: Paternoster, 2009).

[2] Stephen C. Meyer, *Signature in the Cell: DNA and the Evidence for Intelligent Design* (New York: HarperCollins, 2010).

eclectic group of ID proponents is that there are biological entities that are so complex, and so interdependent in their functioning as far as their components are concerned, that their existence cannot be explained by gradual Darwinian processes, pointing to 'design' as the best explanation for their coming-into-being.

The ambiguity of the term 'Intelligent Design' is deepened by the great range of meanings attached to the word 'design', a word commonly used with at least four distinct meanings in the English language. First, design can refer to 'an arrangement of form and appearance, with overtones of purpose'. Any living biological organism will fit such a meaning in a way that rocks and stars do not. It is this meaning that provides the teleological discourse for which biology is famous. Living things exist to feed, survive and reproduce. They are goal-oriented, which is one of the ways in which we distinguish life from non-life.

The second meaning implies a 'specific detailed plan'. This is the language of the engineer or the architect. The bridge is as it is because designed in the mind of the engineer; it could be no other. A third meaning of 'design' refers to the generation of a set of rules, as in chess or a computer game, in which there may be many different outcomes and moves in the game, but in which the designed constraints entail that only some things can happen and not others.

A fourth meaning entails the more general outworking of intentions and purposes. The Government 'designed' the national lottery for the express purpose of funding worthy enterprises in the arts and in sport, in the process extracting money from those least able to afford such expenditure. J.K. Rowling 'designed' seven Harry Potter novels to fulfill her intentions and purposes for the character Harry Potter. Many details of character and of plot changed along the way as the novels were written, but the overall purpose of the series remained intact.

This list by no means exhausts the various nuances of the word 'design', but for our present purposes it is worth noting that in the term 'Intelligent Design' the word 'design' is being used by ID proponents in its second sense to refer to the design of a specific complex entity in biology, akin to the outcome of the design of an engineer or an architect.

The important role that the various meanings of 'design' play in the discourse of ID proponents will become apparent as we consider the main arguments presented in Stephen Meyer's book.

The Signature in the Cell

The author writes this book with a background in the history and philosophy of science and presents his work as a 'trade-book' that surveys the findings of others – biochemists and cell biologists – interpreting the science within the framework of ID arguments. The basic cell biology, of the kind that can be found in any undergraduate textbook, is in general presented accurately and only occasionally does the author's lack of scientific expertise become apparent in the text.

The first part of the book sets the scene with some historical and philosophical material that begins to introduce the author's stance towards the nature of scientific explanations and ideas about information in biology. Several chapters then describe our current scientific understanding of DNA and protein structure, the genetic code, the regulation of gene expression, protein biosynthesis, and the origin of life, underlining the present gaps in our scientific knowledge concerning the incremental, step-by-step processes whereby such entities might come into being. Given the author's perception that these gaps are significant, he concludes this first section by surmising that there might be a "design hypothesis – that could be legitimately considered as a possible explanation for the DNA enigma" (p. 135). The rest of the book is then focused on arguing for the plausibility of such a "design hypothesis".

Meyer did his PhD in Cambridge at a time when the late Peter Lipton was Head of the History and Philosophy of Science Department; it is therefore no surprise to find that his argument depends to a large extent on abductive reasoning involving "inference to the best explanation."[3] In inductive reasoning a universal law or principle is established from repeated observations of the same phenomena. This contrasts with deductive reasoning in which a conclusion is deduced by applying a general law or principle to a new context. In abductive reasoning, however, unseen facts, events or causes in the past are inferred from data available in the present. Evolutionary theorizing, for example, is largely built on abductive reasoning, since the speciation events that led to our present understanding of the evolutionary bush of life are interpretations based on our present knowledge of anatomy, genetics, geographical distribution of species, fossil data, and so forth. We were not there to observe the speciation events; they are inferred from presently known data.

---

[3] Peter Lipton, *Inference to the Best Explanation* (London: Routledge, 1991).

The question for Meyer is: what can we infer as the best explanation for the origins of the kind of biological information and complexity that we see in the structure and functioning of cells? In seven further chapters Meyer then surveys the various attempts that have been made by scientists to explain the origins of this complexity. This involves a discussion of the role of chance and the idea of 'specified information' (Chapters 8-10), the origins of the genetic code (Chapter 11), complexity and emergent properties of systems (Chapter 12), computer models for the origin of life (Chapter 13) and RNA as the original information-containing molecule (Chapter 14). Demonstrating via these seven chapters to the author's own satisfaction that none of these approaches provide adequate explanations, Chapter 15 then argues the case for ID as the 'best explanation' for the origins of specified information, drawing also on ID proponent Bill Dembski's 'design detection' approach to show how this produces the same conclusion (Chapter 16).

Having invoked ID as an 'explanation' for the origins of certain forms of biological complexity, Meyer is only too aware of the kind of critiques that may be mounted against such a claim. Chapter 17 is thus committed to the important question of whether ID counts as a valid explanation for anything, Chapters 18 and 19 to the question of whether ID counts as science, and Chapter 20 to the claim that ID represents a religious belief. A final Epilogue presents a smorgasbord of ideas about ID and an Appendix lists some predictions of ID that, the author suggests, thereby demonstrate that ID has the characteristics of a scientific theory.

In what follows our aim is not to critique all the various points raised in this wide-ranging book, but rather to focus on the main themes that are central to the author's argument.

Is ID science?

A consistent plea within *Signature in the Cell*, and within the ID movement more generally, is that ID be accepted as science rather than as philosophy or theology. Media reports have often suggested that this is merely a political strategy to smuggle religion into US classrooms under the guise of science. However, although there is some evidence to suggest that the attempt to present ID as science is connected with the radical separation of religion and state enshrined in the US constitution, there seems no good reason to accuse ID proponents of bad faith on this point, and the majority, perhaps all, clearly do believe that ID is science. It is therefore a fair question to ask whether the theory suggested in the present volume has the characteristics of a scientific theory.

Philosophers have generally rather given up on the earlier 'demarcation discussion' as to what does or does not count as science. The reason is that it can become a rather sterile discussion with different angels dancing on different pins in different disciplines. It is indeed striking how much science can vary depending on discipline. On the other hand there is no need to go to the other extreme of suggesting that 'anything goes', as in the more anarchic understanding of science presented in the Austrian philosopher Paul Feyerabend's book *Against Method*.[4] And scientists do care a lot about what counts as science, because science as a body of reliable constructed knowledge comes at a high price, with a lot of grant-writing, a hard slog at the bench (in the case of the biological sciences), the sweat of writing papers, peer-review and publication, and the extensive social interactions within the scientific community that play such a key role in the advance of science.

From a practical perspective, it is also vital that scientists get things right. In the case of the biomedical sciences, this may mean the difference between life and death. In the case of genetic engineering, this could make the difference between a hungry population or a nutritionally healthy population. In the case of climate change, this could mean the difference between millions of people being displaced from low-lying areas with the melting of the polar ice-caps, or not. So truth-telling in science is not some academic whimsy, but really does make a difference to our everyday lives, for good or for ill.

Given that fact, it is fairly obvious why scientists really do care about what counts as science. Contra postmodern claims that different language games have equal claims to assent, scientists actually believe that some things are true and other things are false. Somewhat self-serving claims by Meyer that the question of what counts as science represents a "trivial" definitional problem, a "semantic" dispute (p. 399), really won't do – the stakes are much higher than that. Each discipline has its own particular criteria for justifying its truth-claims and Christians of all people should be concerned to be truth tellers within the domain of science, given that the whole purpose of the scientific enterprise is to describe the properties of the created order as accurately as possible.

At a sociological level, one can define science as what is published in scientific journals. Mere publication does not of course guarantee its truth. Often the data are correct but the theory presented with its accompanying interpretations quite false, something that only becomes apparent with the benefit of hindsight. Science is a self-correcting process.

---

[4] Paul Feyerabend, *Against Method* (London: New Left Books, 1975).

The definition of science in terms of what is published in scientific journals only pushes the question further back: what, then, are the characteristics that articles published in scientific journals have in common? And then it becomes rather easy simply to list those characteristics. In practice very few types of science involve the complete list, but there should be a sufficient number of items from the list that no-one from the relevant scientific discipline is in any doubt that a particular piece of investigation belongs to science rather than philosophy, theology, astrology, politics or cooking. The list includes the following:

1. Science in its methodologies excludes questions of ultimate purpose, value and significance. By contrast Aristotelian science included final causality as one of the explanatory features of nature. But empirical science only really began to take off in the sixteenth and seventeenth centuries as it became more modest in its ambitions and began to exclude final causes from its explanations.
2. Science looks for testable hypotheses.
3. Scientific explanations involve problem-solving but do not simply describe the problem using different words; rather, they provide insights into how the physical world actually functions in ways that lead on to further empirical research.
4. The possibility of falsifiability raises the likelihood that an explanation belongs to scientific discourse.
5. Predictive success is the mark of a good scientific theory.
6. Science aims at formulating generalizations about the properties of things whenever possible; the highest levels of generalization are called ‘laws’.
7. Science values mathematics highly and is expressed in mathematical terms whenever appropriate and feasible.
8. Science aims at objectivity and downplays the role of the scientific observer, deliberately excluding the personal.
9. Scientific knowledge aims to be publicly observable and repeatable; it is only taken seriously within the scientific community following publication in peer-reviewed journals.
10. Scientific theories are more valued when they lead to fruitful research programmes involving a growing peer-reviewed literature and, eventually, practical applications.

Here we will pick out just a few items from this list to see how ID fares as science, several of the listed items being discussed at some length in *Signature in the Cell.*

Starting with the last item, we note that since ID was first seriously mooted in the early 1990s it has resulted in a very meager crop of publications in the peer-reviewed scientific literature. ID proponents are quick to point out that there are at least some publications, but in reality the term 'Intelligent Design' or even the word 'design' are absent from such texts. Instead the publication typically seeks to show that a detailed evolutionary description of the origins of the protein folds that provide a particular enzymatic function, or of a complex biological system, or of the Cambrian explosion as revealed in the fossil record is not, it is claimed, explicable by Darwinian mechanisms, implying that some other kind of explanation is necessary. The authors of such papers will invariably claim that they are barred by reviewers and editors from using the language of 'design' in their papers, which might well be the case, but this in turn should help to focus the mind on whether the notion of ID provides any kind of satisfactory alternative explanation, as discussed further below. Given that the notion of 'design' forms no part of such publications, such as those from the experienced molecular biologist Douglas Axe, director of the ID-funded Biologic Institute in Seattle,[5] this point in itself rather underlines the fact that the scientific data published can stand satisfactorily on their own without any need for interpretation along the lines of 'design'.

The criterion of generating a successful biological research programme may readily be illustrated by reference to an actual scientific theory, at first controversial, that was first published in the early 1980s (though with its roots back in the 1960s), suggesting that a mis-folded 'prion' protein was the infectious agent in the disease scrapie that affects sheep and goats. This idea was in sharp contrast to all the other infectious agents that had been discovered to date, be they viruses, bacteria or parasites, that all contain either DNA or RNA in combination with proteins. The idea of a mis-folded protein causing disease by conveying its mis-folded state by a kind of domino effect to other proteins inside the cell, was and is a bizarre notion that remains a topic of active research. But that initial insight in 1982 has been immensely fruitful and it is now known that bovine spongiform encephalopathy in cattle (BSE, "mad cow disease") and Creutzfeldt–Jakob disease (CJD) in humans are likewise caused by prions. The initial characterization of the prion was introduced to the scientific

[5] D.D. Axe, "Estimating the prevalence of protein sequences adopting functional enzyme folds," *Journal of Molecular Biology* 341 (5) (2004):1295-315; D.D. Axe et al., "Stylus: a system for evolutionary experimentation based on a protein/proteome model with non-arbitrary functional constraints," *PLoS One* 3 (6) (2008): e2246.

literature in a single paper,[6] around a decade before ID came to public attention, but since that time there have been 5,396 scientific papers published with 'prion' in their title,[7] illustrating the hugely successful research programme initiated by the prion theory.

The very fact that ID as an idea needs to be defended in trade-books, of which *Signature in the Cell* provides an example, immediately underscores the point that ID has not been found to be a useful idea within the scientific community. A useful scientific idea or theory is quickly picked up by laboratories around the world, so generating a vigorous new programme of well-funded research. ID is not in that category of ideas.

What about testability as a way of assessing whether ID lies within the domain of science? And are there any ways in which it can be falsified? To assess those questions we need to consider a little more carefully the kind of ID explanation that Meyer offers. Cell biology, it is suggested, offers many examples of 'specified information', referring to complex systems such as the 64 triplet codons that specify the sequence of amino acids in proteins. It is the highly specific sequence of amino acids in proteins as specified by the information provided in the genes that gives each protein its unique functional properties. It is on the origins of such complex cellular systems that *Signature in the Cell* focuses.

The next step in the argument depends on the approach, already mentioned, of the 'inference to the best explanation'. Since the language of 'origins' immediately takes us into the realm of the 'historical sciences' such as cosmology and evolution, the only option open to us is to infer what happened in the past by reference to data collected in the present. This then leads on to what the author claims is an "exhaustive" investigation of the various 'materialistic' explanations that have been offered. How satisfactory that "exhaustive" investigation really is will be considered further below, but here we simply note the structure of the argument.

Given the presumed failure of other explanations, Meyer then suggests that the only option left to us is to infer that the complex cellular system in question displays 'intelligent design', "the deliberate choice of a conscious, intelligent agent or person to effect a particular outcome, end, or objective." The argument is made in a similar way to other ID publications by drawing attention to the way in which design is inferred in cryptography, archaeology or in the assumptions underlying the SETI

[6] D.C. Bolton, M.P. McKinley, and S.B. Prusiner, "Identification of a protein that purifies with the scrapie prion," *Science* 218 (no. 4579) (1982):1309-11.

[7] Data from PubMed. Accessed 21 April 2011.

(Search for Extraterrestrial Intelligence) programme. In short, Meyer claims that since we are quite good at inferring the products of intelligent minds, namely human minds because these are the only ones we know about, so, by analogy, we can infer that all this specified information found inside cells is likewise the product of intelligence, and is therefore 'designed'. The way in which the word 'design' is used in this context is in line with its second meaning as summarized in the introduction above, referring to a 'specific detailed plan' as in the language of the engineer or the architect.

Does invoking 'design' count as a scientific explanation? Meyer draws attention to those natural philosophers of previous centuries such as Isaac Newton who did indeed invoke 'design' in their scientific publications. Newton famously maintained that God brought about an occasional 'reformation' in the movement of the planets to correct the irregularities that accrued from the supposed friction occurring as they passed through the 'aether' (an 'aether' which we now know does not exist). Even at the time the German Lutheran philosopher and mathematician Gottfried Leibniz took Newton to task for invoking occasional miracles whereby God would remedy the deficiencies in his creation. Indeed, by the end of the eighteenth century the French mathematicians Laplace and Lagrange had shown that irregularities induced in planetary orbits could be self-correcting. Apparently there was no need for God's occasional corrections.

So the argument that 'design' is in some sense a scientific argument, on the grounds that there have been natural philosophers in past centuries who have invoked it as if it were a scientific explanation, turns out to be a hostage to fortune. The notion in each case has been unfruitful, readily displaced by explanations that refer to physical forces and realities. Indeed, as highlighted in the list given above, it was as early modern science began to abandon the Aristotelian quest for ultimate causes in order to focus on proximate causes that the scientific enterprise really began to make progress.

Invoking 'design' as an explanation for the origins of specified complexity in biology therefore looks like a straightforward category error. A distal Aristotelian cause is being invoked to explain what really requires a better set of proximal explanations to give insights into the processes whereby complex biological systems come into being. Since the problem is one concerned with molecules, their properties, and their coming into being in certain kinds of assemblies, then we expect a scientific explanation that will address how this might occur. The problem with invoking 'design' as an explanation becomes particularly acute when we consider the question: "What further experiments should I now do based on this putative explanation?" The answer is that there aren't any. The situation is reminiscent of Peter Woit's critique of the proliferation of speculative multiverse models as exemplified by the myriad solutions of string theory. His book was entitled *Not Even Wrong*.[8]

It is this point that explains why the ID notion has not led to any kind of coherent research programme. The most it can do is to motivate research projects showing that explaining the origins of complex biological entities, like protein folds, is a challenging problem. But we knew that already; the notion of ID does not add anything to our absence of knowledge in this research area, except to highlight the challenge. Furthermore, there are laboratories all round the world working on scientific challenges, such as the evolutionary origins of protein folds, and making significant progress without any need to invoke the notion of 'design'.[9]

Finding tests that would falsify the idea of 'design' as an inference to the best explanation for the origins of biological complexity is as problematic as understanding how 'design' counts as an explanation in the first place. Describing a complex system as 'designed' sounds more like adding an extra descriptive label than like a scientific explanation. So what experiment would falsify such a label? Meyer offers not a single possible experiment that would help in this respect. Instead he offers philosophical arguments that are presented as a type of 'testing' (pp. 404-407) and as a type of 'falsifiability' (pp. 428-433), but which in reality are not.

The misunderstanding of the author concerning what counts as scientific testing, certainly in the arena of the biological sciences, is revealed in this comment: "Since empirical considerations provide grounds for rejecting historical scientific theories or preferring one theory over another, such theories are clearly testable. Like other historical scientific theories, intelligent design makes claims about the cause of past

---

[8] Peter Woit, *Not Even Wrong: the Failure of String Theory and the Search for Unity in Physical Law* (New York: Basic Books, 2006).

[9] See K.B. Zeldovich et al., "A first-principles model of early evolution: emergence of gene families, species, and preferred protein folds," *PLoS Comput Biol* 3 (7) (2007): e139; Y. Lin et al., "Engineering of functional replication protein a homologs based on insights into the evolution of oligonucleotide/ oligosaccharide-binding folds," *Journal of Bacteriology* 190 (17) (2008): 5766-80; E. Ferrada, and A. Wagner, "Evolutionary innovations and the organization of protein functions in genotype space," *PLoS One* 5 (11) (2010): e14172; also Dennis Venema. "Evolution and the origin of biological information," http://biologos.org/blog/evolution-and-the-origin-of-biological-information-part-4/. Accessed 24 April 2011.

events, thus making it testable against our knowledge of cause and effect." The author then goes on to claim that "*historical* scientific theories typically do not make predictions that can be tested under controlled laboratory conditions." But this is not the case. For example, natural selection can readily be shown to operate under controlled laboratory conditions, as can changes in reproductive fitness in controlled population experiments. Furthermore, evolutionary theory could readily be refuted by an accumulation of counter-evidence. Rabbit fossils could be found in the Precambrian. Human footprints could be preserved along with those of the dinosaurs in the mud from the same riverbed. Had the genetic code been different for all the different families of life, then the united Darwinian tree of life would have split into several different trunks with the inference that there were several independent origins for the first cells (in reality all living things examined so far share essentially the same genetic code). There are many other examples.

By contrast, Meyer's arguments about testing and falsifiability essentially boil down to different ways of re-phrasing the notion that 'design' provides the inference to the best explanation for the origins of biological complexity. 'Testing' in the author's vocabulary refers to 'testing' the validity of one historical inference against another, and 'refutation' becomes the possibility that a much better inference could be found than that provided by 'design'. But this is not what these terms mean in biological theories, historical or otherwise: they mean the ability to do actual experiments that will count one way or another in the assessment of a theory. Evolutionary theory suggests actual experiments that can be carried out to show how novel information comes into being, as described further below. But the notion of 'design' proposed in *Signature in the Cell* is so nebulous that it explains everything and nothing at the same time.

This point is well illustrated by Appendix A of the book (pp. 481-497) headed 'Some Predictions of Intelligent Design'. The problem with these so-called predictions is that they are generally so vague that virtually any outcome could be lauded with the benefit of hindsight as a 'successful prediction'. For example, Bill Dembski's 1998 prediction is cited as an exemplar: "On an evolutionary view we expect a lot of useless DNA. If, on the other hand, organisms are designed, we expect DNA, as much as possible, to exhibit function" (p. 407). As it happens some organisms have a very high percentage of their DNA committed to protein-encoding genes, whereas in the case of our own genomes only 1.5% is designated for protein-encoding genes, another 3.5% or so to known regulatory elements, a massive 46% to transposons (segments of "copy-and-paste"

DNA, sometimes known as "jumping genes" or "mobile genetic elements") and around 8% to retroviral insertions. The known functions of our genomes are currently restricted to the 5% that contain protein-encoding genes together with their regulatory sequences, although there is good evidence that other segments within the remaining 95% are functional in various ways. But the highly repetitive nature of much of that 95% renders it very unlikely that all of it will turn out to have a function. So what? It's not clear from the perspective of a 'design' argument why the outcome should matter either way. The human genome clearly does a very good job at encoding human bodies, just as genomes of other organisms that have much higher proportions of protein-encoding genes do a very good job at encoding those organisms as well. There are no doubt some good reasons why genomes are organized in different ways and the rapid pace of discovery should shed light on this in the coming years. For the present it is worth noting that the human genome gives us really brilliant insights into our evolutionary past,[10] providing by far the best evidence that we have of our common descent with the apes. Those insights are real, but the question of whether or not the human genome displays 'design' is scientifically sterile, leading to no unambiguous predictions, interpretations or new experiments.

Much of the justification for 'design' as an inference to the best explanation in *Signature in the Cell* comes from the supposed inability of other theories to explain the origins of biological complexity adequately. As Meyer comments in proposing ID as a rival 'scientific theory': "Despite the thorough search described in Chapters 8-14, I found no other causally adequate explanations for the DNA enigma" (p. 330). Given that the "DNA enigma" (meaning the origins of the genetic code) allied with other complex cellular systems do indeed represent significant arenas of scientific ignorance, although not quite as much as the author suggests, as we will consider further below, the most honest statement is simply to say: "At present we don't know how such-and-such an entity came into being, although we have some speculative ideas." Science is about problem solving and there is nothing that scientists like better than a really big problem to solve. But science is also about the "art of the soluble,"[11] and picking an appropriate and achievable problem to solve is also a key to a laboratory's success.

Darwin had little idea in 1859 as to how inheritance occurred at the

---

[10] Denis R. Alexander, *Creation or Evolution - Do We Have to Choose?* (Oxford: Monarch, 2008); Denis R. Alexander, *The Language of Genetics - an Introduction* (Philadelphia: Templeton Foundation Press, 2011).

[11] Peter B. Medawar, *The Art of the Soluble* (London: Methuen, 1968).

molecular level and the ideas he did have turned out to be wrong. In fact it took a whole further century before that molecular information became available. It's just as well that the scientists of Darwin's day didn't just throw their hands up in despair and say that "inheritance is all so complex and difficult that it must be designed," otherwise we wouldn't have the wonderful insights into the processes involved that we have today. Likewise it might well be the case that it takes another century or so to gain a complete understanding of the processes whereby the complex biological systems described in *Signature in the Cell* come into being. Simply labeling them 'designed' seems unlikely to help in that scientific voyage of exploration.

The philosophical background to ID

It might seem odd that a theory presented as science should be propagated via books, such as this one, written for the general reading public, as well as via films and the media. As the author points out, the theory of natural selection was also first presented in book form in Darwin's *On the Origin of Species*. But there the similarity ends. Darwin packed his book with scientific data, much of it collected himself during his voyage on *HMS Beagle*. Furthermore, in 1859 science was less professionalized than it was to become a few decades later and the pressure to publish in peer-reviewed scientific journals was much less. Today that pressure is insistent and in the regular UK University and Research Institute assessment exercises that decide their funding allocations, books written by scientists are of no consequence, the only items that count as research outputs being peer-reviewed articles in scientific journals.

In contrast to scientific literature, *Signature in the Cell* also presents its theory in quite a polemical way as if much hangs on it, much more indeed than merely increased knowledge about the way that cellular processes come into being. Indeed, reading the book was somewhat reminiscent of reading the Appendix to Daniel Dennett's book *Breaking the Spell* in which Dennett mounts a vigorous defence of the notion of the 'meme', a supposed unit of cultural inheritance.[12] If the value of a scientific metaphor has to be defended in a popular book, then you know that it must be of little use to the scientific community for the simple reason that useful metaphors that do real work in the laboratory are quickly taken up spontaneously and used in any case by the community. Recourse to

[12] Daniel C. Dennett, *Breaking the Spell* (London: Penguin, 2007).

defence in popular books is a sure sign that the cause for the favoured metaphor is already lost.

The reasons for the polemical fervor in the literature of ID proponents, which includes *Signature in the Cell*, may be readily understood by the realization that Darwinian evolution, which is here the object of attack, is no mere scientific theory in the eyes of its author, but rather comes loaded with all kinds of philosophical baggage. Darwinian evolution is perceived as due to "wholly undirected processes such as natural selection and random mutations" which "alone can produce the intricate design-like structures in living systems" (p. 4). ID theory "does not challenge the idea of evolution as change over time or even common ancestry, but it does dispute the Darwinian idea that the cause of all biological change is wholly blind and undirected" (p. 4). On page 37 Meyer makes the important distinction between idealism (mind is the prime or ultimate reality), of which theism is a species, and naturalism, which maintains that matter is the ultimate reality. Whereas the definition of this distinction is unremarkable, what is contentious is the author's habit in the pages that follow then to describe as "naturalistic" what most people would simply refer to as scientific theories. For example we read that "Darwin sketched out a purely naturalistic scenario for the origin of life" (p. 43). The emergence of early ideas on the chemical evolution of life in the 1950s is referred to as part of a "grand materialistic story" and as a "naturalistic account" (p. 57). The contrast made with ID is revealing: "what natural selection and mutation lack, intelligent selection – purposive or goal-directed design – provides" (p. 336). It is the "inadequacy of proposed materialistic causes" that provides part of the basis for the author's ID argument (p. 376).

In referring to scientific theories as *ipso facto* 'naturalistic' the author contradicts his own historical sections of *Signature in the Cell* in which he points out that many of the early modern natural philosophers who established modern science were people of deep Christian faith who saw the whole universe as God's good creation. Given that naturalism refers to the "view of the world that excludes the supernatural or spiritual" (Oxford Dictionary), it is therefore difficult to see how a theist could describe any of its properties as being "naturalistic". If we are existing within God's creation, then surely all scientific descriptions without exception must in some sense be descriptions of God's handiwork, an outworking of his will and purpose. Yet within the author's philosophical paradigm it seems that the universe is divided into 'naturalistic' and presumably 'non-naturalistic' domains, and it is the origins of these latter domains that the author wishes to explain by ID. We should not lose sight of the profound philosophical and theological ambiguity introduced by this attempt to divide the world into such radically different categories.

A similar problem arises with the description of straightforward scientific theories as "materialistic," a word clearly used, as in the quotations cited above, with negative connotations. Indeed this is a loaded word. Christians, for example, are warned against the dangers of materialism, of becoming materialistic in their thinking, referring to the tendency to accumulate wealth on this earth and not lay up treasure in heaven. The problem in using this word to refer to science is, of course, that the negative connotations are carried over, quite unnecessarily, into that domain. Yet Christians of all people have a high view of the value of the created order. It has often been said that Christianity is a very 'materialistic' religion in this sense. God loves materials because he has made so many of them. In Genesis 1 the material world that God creates is continually declared to be 'good'. Christians should rejoice in the privilege of being able to explore God's material world through the techniques and approaches that science provides. Labeling scientific theories that describe God's materials as 'materialistic' is a misuse of language.

Where does biological information come from?

Much of *Signature in the Cell* is given over to a discussion of the origins of biological information in general and to the origin of life in particular. The two go hand-in-hand. Without specified complexity living organisms would be unable to organize themselves, to reproduce and to pass on their properties to their progeny, the defining properties of 'life'. It is information that renders a cell very different from a stone. Where does it come from?

Curiously Meyer dismisses natural selection as the source of new information without further discussion. Nor is the distinction made that most biologists make between classical neo-Darwinian processes involving genetic variation plus natural selection, and the very different kind of science involved in questions relating to the origin of life. In its strict sense Darwinian evolution can only occur when DNA exists that then varies, with differential reproductive success determining the chance of particular sets of beneficial variants being represented in subsequent generations. Even bacterial evolution is sometimes referred to as being "below the Darwinian horizon" (a phrase of Carl Woese) given the prevalence of bacterial lateral gene transfer that subverts merely 'vertical' processes for the transfer of information. And when it comes to the origin of life, with the challenge of explaining the origins of the first information-containing molecules, then more relevant than Darwinian theory *sensu stricto* is the work of chemists, biophysicists, biochemists and geophysicists. It is no accident that much of the current funding for origin of life research focuses on such disciplines and comes from NASA with its interest in finding life on other planets.

Genetic variation coupled to natural selection provides a constant flow of new information into populations of living organisms. For a trivial example, we only need to look around us at our fellow humans in a crowded street to see hundreds of examples of differential information before our eyes. The tall and the short, the curly-haired and the straight-haired, the blue-eyed and the brown-eyed, the fair skinned and the dark skinned, the small footed and the large footed – all these different features provide visual exemplars of the generation of new information in our inherited genomes, that encode such diversity, over the past millions of years. Some of the diversity is due to neutral mutations that are not subject to natural selection, but other differences are clearly selected for, depending on environmental context. It is no accident that East Africans make such great marathon runners. The short, stocky legs of the Inuit contrast with the long lanky legs of the Kenyan, the former adapted to retain heat, the latter adapted to rapid heat-loss and the challenge of chasing game for long distances over the Serengeti. Neanderthals coping with the European ice-ages likewise developed shorter stocky legs as an adaptation to the cold.

For more dramatic examples of the acquisition of new biological information one only needs to walk out into the local park, countryside, or back garden with eyes open, and admire the huge diversity of life-forms in both the plant and animal kingdoms. As Meyer points out, many ID proponents believe in common descent, so the thousands of different forms of variant genome that underlie the rich collection of species before ones eyes are all illustrations of the ways in which novel genes or variant forms of known genes undergo natural selection to create the varied array. New information appearing in evolutionary biology is not some esoteric notion that requires intensive study of information theory to appreciate; it is there for all to see every time we look out of the window.

Of course when we look at biological diversity in this way we are seeing the outcome of millions of years of evolution. Since the process of acquiring new information generally takes many generations, the acquisition is most readily investigated in the laboratory in organisms with rapid replication times, such as bacteria which can divide every 20 minutes under the right conditions. Rich Lenski and colleagues, then at the University of California, started an experiment of this type on 24 February

1988. They started growing a series of twelve populations of the bacterium *Escherichia coli*, all derived from a single bacterium and fed using glucose. The evolution of different strains of these bacteria from the original parental cells has now been tracked for a period of more than twenty years.[13] Each day about half a billion new bacteria grow in each flask, involving the replication of the same number of bacterial genomes, and in total about a million mutations occur in each flask as the bacteria divide. Since there are only about 5 million base-pairs in the bacterial genome, this means that every few days virtually the whole genome will be subject to genetic analysis to see whether any of the new mutations might be useful. In practice the vast majority are not, but new mutations occasionally come along that provide some growth advantages.

Every night the bacteria run out of their glucose food source and become dormant, so bacteria that cope best with this changing environment have a big advantage. The next day about 1 percent of the culture from each flask is used to start a new culture with a new supply of glucose. Most of the beneficial mutations that occur provide up to a 10 percent growth advantage, and such mutations spread rapidly through the population as the progeny carrying the mutation have this modest growth advantage. What Lenski found was that the evolution of the different flasks of bacteria, as measured by their growth, developed not in a smooth trajectory but in a series of abrupt jumps as advantageous mutations took over the population.

After more than a decade of sub-culturing the twelve flasks, something rather extraordinary happened at generation 33,127. One of the cultures "discovered" how to use citrate – a chemical used to stabilize the pH and so present in all the flasks since the beginning – as a food source. This gave this population a huge growth advantage as it was no longer dependent upon glucose as a food source. Further analysis revealed that the capacity to use citrate could not evolve all in one step, but took three different mutations to achieve. The two "background" mutations had to occur first, and the third critical mutation then enabled the complete ensemble of three mutations to allow the use of citrate, thereby opening up a whole new way of living for the colony. Such experiments provide a striking example of the way in which new information can be generated by the Darwinian process of genetic variation plus natural selection.

Much evolution occurs by the process of gene duplication. Normally during DNA replication and cell division the same number of genes are

---

[13] J.E. Barrick et al., "Genome evolution and adaptation in a long-term experiment with Escherichia coli," *Nature* 461 (7268) (2009):1243-7.

passed on to the daughter cells. But occasionally a segment of DNA is duplicated twice and passed on in the germ line cells, and occasionally this segment may contain one or more genes. Gene duplication provides an important way of generating new information during the course of evolution because, with time, the duplicated gene can accumulate other types of mutation that provide it with new functions, different from those of its original parental gene. An interesting example of this genre involving receptors and their ligands may be found in an on-line article by Dennis Venema, a biologist from Trinity Western University in British Columbia.[14]

The idea that new information cannot arise from normal Darwinian processes is fallacious and there are hundreds of detailed examples in the scientific literature showing how this has in fact occurred. A very similar process involving the generation of new information by random genetic variation followed by selection occurs in our bodies every time we fight of a foreign invader, known as an 'antigen' (for example those coming from bacteria or from noxious chemicals). This process takes place in the antibody-making cells of our bodies, known as B cells. Through a process of random gene shuffling a vast array of B cell antibodies are produced so that each B cell has specificity to a particular antigen. When the antigen invades it binds to that tiny population of B cells that recognize it, leading to their selective reproduction. This is accompanied by further random gene mutation and selection until antibodies are produced that have much higher affinity for the antigen than in the first wave of recognition and binding. Genuinely new information has been produced by random variation and selection. And this is going on in our bodies all the time.

The origin of life

The particular focus of *Signature in the Cell* concerns the origins of the biological information required to operate the basic processes of cell biology, including the genetic code and the molecular mechanisms required to convert the information encoded in DNA into the specific amino acid sequences of proteins. Certainly this is a tougher problem to crack than the question of how new biological information is generated once the genetic code has been established. One of the reasons that it is so tough is the immense time-scale involved. There is tentative evidence for

---

[14] Dennis Venema. http://biologos.org/blog/evolution-and-the-origin-of-biological-information-part-3-csi-on-steroids/. Accessed 25 April 2011.

life by 3.8 billion years ago and firm evidence for cellular life from the period 3.0-3.5 billion years ago. This means that the basic molecular mechanisms of the cell have remained virtually unchanged for the past 3 billion years or more and it is really hard to see from this great time distance how those mechanisms may have come into being.

Space does not allow a thorough critique of Meyer's survey of the various theories that have been broached since the 1950s to elucidate the origins of the first cells. Suffice it to say that the survey focuses largely on historical theories that have not got very far (described in 123 pages) and very little attention is given to the contemporary state-of-play (only 28 pages). This gives the false impression that very little progress has been made in this field. It is certainly the case that there is as yet no convincing series of molecular steps leading from non-life to life. But on the other hand more progress has been made than the author's account suggests. Furthermore, there is little discussion concerning the several current theories about the origins of the genetic code, and some of the information that the author does provide on this point is wrong.

For example, Meyer suggests that the chemical binding of any given amino acid to a transfer RNA (tRNA) is the same, and it is only the enzymatic assistance provided by the aminoacyl-tRNA synthetases that ensures that the correct amino acid binds to its specified tRNA (p. 248). Although the synthetase enzymes are indeed now crucial in this respect, it is interesting to note that preferential direct binding of some amino acids to their cognate tRNAs does indeed occur, even in the absence of help from a synthetase enzyme, and it is this which has given rise to the so-called 'Escaped Triplet Theory' which proposes that it was amino acids that originally selected the primitive genetic code.[15]

To appreciate this discussion one needs to know that each of the twenty amino acids that make up proteins has its own designated tRNA. Each tRNA includes an acceptor 'arm' that attaches by a chemical bond to a specific amino acid. At the other end of the tRNA molecule is another arm, right in the middle of which is a triplet sequence of nucleotides known as the anticodon. The anticodon corresponds exactly to the codon in the messenger RNA (mRNA) sequence, just as a plug fits into a complementary socket, and the sequence of mRNA triplet codons in turn is specified directly from the DNA of the cell. In this way the genetic information of the DNA is 'translated' into a specific sequence of amino acids during protein biosynthesis.

---

[15] M. Yarus, J. G. Caporaso, and R. Knight, "Origins of the genetic code: the escaped triplet theory," *Annual Review of Biochemistry* 74 (2005):179-98.

The 'Escaped Triplet Theory' is based on the fact that if we take a large number of randomised RNA sequences and determine experimentally which ones interact preferentially with a certain amino acid, then we are far more likely than mere chance would allow to pick out precisely those RNA molecules that contain the triplet codon, and in some cases the anticodon as well, that we know specifies that particular amino acid. In other words, the 3D structure of the RNA molecule is influenced to a considerable degree by the presence or absence of a particular sequence of three nucleotide bases. This amino acid tends to bind to those RNA molecules that have certain triplet bases and not others. For example, in one randomised binding study, the amino acids arginine, isoleucine and tyrosine bound preferentially to RNA molecules containing either their known codons or anticodons.[16] In the present translation system, as used by all cells today, such differential binding affinities are irrelevant because the tRNAs act as adaptor molecules to bring each amino acid into its correct position. The idea in the 'Escaped Triplet Theory' is that we are seeing some molecular characteristics that are remnants of an earlier more primitive system for synthesising proteins using RNA as a template. The greater affinities of a sub-set of amino acids for RNA molecules containing precisely the 'correct' codons and/or anti-codons does seem to require some kind of explanation, and an 'inference to the best explanation' points to a role in an evolutionarily early coding system.

Unfortunately *Signature in the Cell* contains no discussion of the 'Escaped Triplet Theory', nor of the very substantial scientific literature on the origins of the genetic code. Why is this particular set of 64 triplet codons used and not another? There appear to be some very good chemical and evolutionary reasons why this should be the case, reasons that lie beyond the scope of this chapter to survey. If and when life is discovered on other planets, it would not be surprising to discover that the triplet codon system for exoplanetary life looks similar to the one with which we are familiar here on planet earth.

Meyer considers the 'RNA-world hypothesis' for the origin of the first self-replicating system in a rather brief chapter. This is the idea, important in contemporary origin of life research, that the first information-containing molecule was RNA rather than DNA. RNA has the same nucleotide base structure as DNA except that uracil is used in place of thymine. One of the reasons for preferring RNA for life's origin is because certain types of RNA molecule, known as ribozymes, can act as enzyme

---

[16] A.S. Rodin, E. Szathmary, and S. N. Rodin, "On origin of genetic code and tRNA before translation," *Biology Direct* 6 (2011): 14.

catalysts as well as replicators, thereby potentially getting around the chicken-and-egg situation posed by the need for a molecule to replicate itself in the absence of a protein enzyme catalyst. Today the task of replication of both DNA and RNA is helped along by polymerase protein enzymes, but if RNA came before proteins in the origin of life then clearly something else would need to act as the catalyst. In the RNA-world idea, that 'something' is RNA, and it is now known that ribozymes can catalyse all the different steps needed for a self-replicating system, albeit with efficiencies much less than the protein enzymes that cells now use.

Meyer briefly reviews some of the difficulties with the RNA world hypothesis. Certainly there are plenty of challenges for the theory, as there are for all origin of life scenarios. But his chapter on the 'RNA world' hypothesis well illustrates the dangers of placing one's argument upon current weaknesses in any particular theory. Science has a habit of speeding ahead very fast. Even in the brief time since *Signature in the Cell* was published, further results have been accumulating that render some of Meyer's critiques less potent.

For example, the author points out that the ribozymes engineered with polymerase activity are highly limited in their actions. But in 2011 a paper appeared describing a more sophisticated RNA polymerase ribozyme able to synthesise RNAs of up to 95 nucleotides in length.[17] Furthermore, using this new ribozyme, an RNA was synthesised which itself had a different enzymatic activity. These types of 'proof-of-principle' experiments are important in establishing that such molecules can exist with these particular functions. There are currently about 50 new papers being published every year in the scientific literature with the word 'ribozyme' in their title, and many other papers beside which report new findings on this burgeoning field. It would be unwise to predict at this stage just how many properties these fascinating molecules might be able to demonstrate.

The author also suggests that protein synthesis requires a transition from the proposed RNA-based early world to the 'modern' system that uses protein enzymes. But this is not actually true. The ribosomal RNA that plays the key role in today's cells to catalyse the formation of the peptide bonds that link amino acids to make proteins is itself a ribozyme – an RNA enzyme and not a protein enzyme.[18] Proteins are used to stabilise and direct peptide bond formation, but the enzymatic role is mediated by

---

[17] A. Wochner et al., "Ribozyme-catalyzed transcription of an active ribozyme," *Science* 332 (6026) (2011): 209-12.

[18] V. Ramakrishnan, "What we have learned from ribosome structures," *Biochem Soc Trans* 36 (Pt 4) (2008): 567-74.

RNA. So in this case the earlier putative RNA world, in contrast to Meyer's claim, never made the transition to a protein enzyme.

Origin of life research is tough and even if a completely coherent and convincing set of molecular steps can be worked out in the laboratory, showing how living reproducing entities can come into being starting with basic chemicals, we could not be sure that this is how life actually began. The author asks: "Are there physical or chemical forces that make the production of information-rich molecules inevitable under plausible prebiotic conditions?" (p. 232). That seems quite likely, but it is not clear why the comment helps the author's argument. If true, then equally it provides fresh impetus to those working in the origin of life research area who are seeking to understand the emergence of the first replicating entities through the normal processes of physics and chemistry.

Christian creation theology and ID

My first meeting with the author of *Signature in the Cell* was at a barbeque in the back garden of friends in Cambridge. We soon got down to a vigorous discussion about evolution. This was my very first exposure to the ideas of what is now known as ID. The term was quite unknown at the time. I was very soon struck by the thought that this sounded very like the old god-of-the-gaps idea only dressed up in new clothing. My suggestion of this possibility was not greeted warmly. But *Signature in the Cell* does seem to represent essentially the same idea, except that it may be more accurate to label it 'designer-of-the-gaps' in this case. *Plus ça change, plus c'est la même chose.*

There is little overt theology in *Signature in the Cell* (though the author does occasionally promote a strong form of natural theology) and the theological implications of the book's main thesis actually run strongly counter to the traditional creation theology espoused by the Abrahamic faiths. The author states that "Though the theory of intelligent design does not identify the agent responsible for the information – the signature – in the cell, it does affirm that the ultimate cause of life is personal" (p. 450). All theists will agree that the ultimate source of everything that exists is personal, namely, God. However, the attempt to localize the actions of the "creative intelligence" to particular aspects of the created order, specifically those aspects that science is presently not very good at understanding, is deeply problematic.

The author makes a strenuous attempt to claim both that current science does not explain the origins of cellular complexity and that this will *never* be the case even in principle. For example, in discussing the present inability of science to provide a complete understanding of the molecular processes describing the origin of life, the author claims that "it is probably more accurate to characterize this supposed "absence of knowledge" as knowledge of absence, "since it derives from a thorough search for alternative materialistic causes and a thorough evaluation of the results of numerous experiments performed over several decades" (p. 376). But this is a hostage to fortune. In science we learn to say "never say never." As already illustrated, today's gap in our knowledge is tomorrow's paper in *Nature* or *Science*. The problem with a "thorough search" is that it can only apply to this particular moment in the history of science, which is a very long history indeed, with hopefully much further to run. And as already indicated, for example with theories about the origin of the genetic code and the 'RNA world' idea for the origin of life, the author's search, even on the recent scientific literature, does not seem to have been that "thorough."

Further problems come with the author's claim that "New laws will *never* explain the origin of information, because the processes that laws describe necessarily lack the complexity that informative sequences require" (p. 268, author's italics). Again, this claim wishes to carve out a permanent gap for the intelligent designer, but the claim itself is quite fallacious. New information is being generated all the time in billions of genomes in organisms across the planet, a generation of information well described via our current understanding of physics, chemistry and biology. Just because we currently have no detailed explanation of how the first RNA or DNA molecules came into being is no guarantee that we will not understand those origins much better in the next few years, or a century's time, or two centuries' time – the time-scale is not that important because even a century or so is a mere drop in the history of human acquisition of knowledge about the world. One of the curious features shared by ID proponents is their tendency to treat current science as if it were some static body of knowledge, when in reality millions of peer-reviewed scientific papers are pouring off the press each year. Science is an incredibly dynamic and fast-moving enterprise.

For those holding to the traditional Christian understanding of God as creator, the question of whether or not we are currently able to understand some component of the created order in scientific terms is theologically irrelevant. Within a theistic world-view, God is the ground and source of all existence. All the materials and energy of the universe have the properties that they possess due to God's faithfulness in sustaining and upholding the created order moment by moment. If there is a multiverse, then the same theological point applies to the multiverse; it makes no difference. Christians, of all people, have no hidden investments in scientific ignorance.

This emphasis in Christian theology is often referred to as the 'immanence' of God in creation. As the prologue to John's Gospel has it: "Through him all things were made; without him *nothing* was made that has been made" (John 1:3, New International Version; my italic). In Colossians Chapter 1, Paul speaks of the Son being "the image of the invisible God, the firstborn over all creation" (verse 15), and then writes: "For by him all things were created: things in heaven and on earth, visible and invisible, whether thrones or powers or rulers or authorities; all things were created by him and for him. He is before all things, *and in him all things hold together*" (verses 16–17, my italics). The point is further underlined by the writer to the Hebrews when he writes that "The Son is the radiance of God's glory and the exact representation of his being, *sustaining all things by his powerful word*" (Hebrews 1:3, my italics). God is the one "for whom and through whom everything exists" (Hebrews 2:10). If God did not keep on willing the created order to exist by his powerful Word, then it would stop existing. Creation is about ontology: why does something exist rather than nothing?

Calvin wrote about the immanence of God in the created order in this way: "For it is the Spirit who, everywhere diffused, sustains all things, causes them to grow, and quickens them in heaven and on earth. ... In transfusing into all things his energy, and breathing into them essence, life and movement, he is indeed plainly divine" (*Institutions* 1, 13-14).

Once we grasp the biblical insistence on the immanence of God in the created order, then we will be wary of the use of descriptions of the created order as being 'naturalistic', as already discussed. It is clear that those who disavow that particular philosophy will not wish to refer to scientific descriptions as being 'naturalistic'. In an attempt to address this problem, it is common to refer to 'methodological naturalism', meaning that there is no need to bring theological language into our scientific descriptions of the world. Whilst that point is perfectly valid, it might be simpler, in order to avoid misunderstanding, to drop the redundant phrase, and simply speak of our 'scientific' descriptions of the world.

The rhetorical identification in *Signature in the Cell* of 'scientific explanations' with 'naturalistic explanations' helps to explain the angst that is always present in the writings of ID proponents. For them it is clear that scientific explanations carry overtones of autonomy, as if once the clouds of scientific ignorance have been blown away, then this represents a triumph for naturalism. For the Christian, however, nothing could be further from the truth. All increases in our well-validated understandings of the created order are causes for thanksgiving and worship to the God who is the creator of all.

Concluding comments

Signature in the Cell is but one of a cluster of books by ID proponents that present essentially the same arguments, albeit using a range of different examples. It is difficult to see how the ID position can be developed any further, given that its core idea depends on our present scientific ignorance concerning the origins of various complex biological processes. As the cloud of ignorance gradually dissipates with the passage of time, one can only imagine that the central ID argument will in turn likewise gradually fade away. But, for the moment, the investment of evolution with the ideological rhetoric of atheism by a small albeit vocal sub-set of biologists will no doubt continue to provide the cultural and political energy that the ID movement needs to keep going.

In the interim one can only express regret that so many sharp minds, acting from sincerely held convictions, can expend so much time and energy on what can only be described as a distracting ‘red herring’ as far as Christian faith is concerned. There are, however, some grounds for optimism. If Christian ID proponents, such as the author of *Signature in the Cell*, could be encouraged to adopt the kind of robust theism that characterizes the traditional Christian understanding of creation, then it is not impossible that the ID movement, perhaps by small evolutionary increments, might eventually merge with the Christian mainstream, which sees the intelligibility of the universe as consistent with a God who has designed all of it with particular goals and purposes in mind.

CHAPTER NINE

A SEMIOTIC THEOLOGY OF NATURE

ANDREW ROBINSON
AND CHRISTOPHER SOUTHGATE

The conference at which this paper was presented celebrated the 150th anniversary of the publication of Charles Darwin's *Origin of Species* and the 200th anniversary of his birth. We wish to introduce a theological approach to Darwinian evolutionary biology based on the work of another Victorian Charles – Charles S. Peirce.

Both men have been described as geniuses, but their genius was of very different kinds. Charles Darwin applied himself to a detailed and sustained study of living and extinct species and their relationships, and set out compelling evidence that all species are related by common descent. In addition, he proposed a mechanism for the process of evolution, namely, natural selection. Evolution was already 'in the air' when Darwin was developing his theory, and he was forced into publication when he became aware that someone else (Alfred Russel Wallace) had independently hit upon the concept of natural selection. Darwin's genius arguably lay in his single brilliant insight that natural selection plus descent with modification could explain the transmutation of species. It was his single-minded pursuit and testing of this hypothesis in diverse fields, however, that translated the brilliant insight into an intellectual revolution of which he is rightly regarded as the founder. Those of us for whom Darwin is a hero might wish to add that his genius was matched by admirable personal qualities as a loyal friend, devoted husband, and loving father.

Charles Peirce was an altogether different kettle of fish. His personal life was chaotic, he was an unfaithful and possibly violent husband, and he made so many enemies that he never held a permanent academic position. Whereas the *Origin of Species* was an immediate best-seller, Peirce never managed to write a book of his own. His published writings are often regarded as difficult and inconsistent, and the task of making available his unpublished work is a long way from completion (six volumes of collected works have been completed by the Peirce Edition Project, out of a projected total of 30 volumes). Nevertheless, there seems to be increasing interest in Peirce's work in a variety of fields. Darwin's genius resulted in an insight that was readily grasped. As T.H. Huxley famously chided himself: "how extremely stupid not to have thought of that."[1] In contrast, Peirce's genius was to see the world in a way for which the world was not yet prepared. In this chapter we will be outlining some of the key features of our joint exploration of how a Peircean perspective may illuminate the field of science and religion, specifically the relation between evolutionary biology and Christian theology.

Evolution, biosemiotics and the Trinity

Our original interest in Peirce arose from our attempts to articulate Christian theology in a way that takes with utmost seriousness the findings of evolutionary biology. For example, is it still possible to preserve some sense of human *distinctiveness* – as reflected in the way that Judaeo-Christian theology has traditionally wanted to speak of humans as distinctively created in the image of God – given that Darwinism suggests a continuity of relationship between all living things? How do we understand the relationship between human beings and the rest of the living world, stretching back through the deep-time of evolutionary history? Second, what kind of theological anthropology is possible, given the apparent sufficiency of naturalistic explanations of biological evolution in general and of human evolution in particular. Third, what is the theological significance of the chanciness of the evolutionary process: is life in general and humanity in particular a purely contingent occurrence, or are there generic aspects of evolutionary outcomes that would be expected to be repeated in any re-running of the evolutionary 'tape'?[2]

The field of *biosemiotics* caught our eye because it seemed to offer a new way of thinking about such questions. Semiotics is the philosophical

---

[1] Leonard Huxley, *The Life and Letters of Thomas Henry Huxley*, Volume 1 (London: Macmillan, 1900), 189.

[2] The way in which the 'semiotic model' outlined here may be understood as a response to these issues of continuity, naturalism and contingency is discussed in Andrew Robinson, "Continuity, Naturalism and Contingency: A Theology of Evolution Drawing on the Semiotics of C.S. Peirce and Trinitarian Thought," *Zygon* 39 (2004), and in Andrew Robinson, *God and the World of Signs: Trinity, Evolution and the Metaphysical Semiotics of C.S. Peirce* (Leiden: Brill, 2010).

field concerned with the structure and function of signs and sign-processes. Biosemiotics is the application of semiotic approaches to biology. According to biosemiotics, sign-processes – the processes of representation and interpretation – are a fundamental feature of life, a thread connecting all living things to one another. The genetic 'code' is one aspect of this though, as we shall suggest below, the capacity of living organisms to 'interpret' their environments may be even more fundamental and ancient than the code. According to the biosemiotic perspective, then, the continuity of evolution might be understood as a continuity and growth of *semiotic* capabilities, culminating in what Terrence Deacon has called 'the symbolic species'.[3] From this perspective humans may be regarded as genuinely distinctive, and yet our distinctiveness is in continuity with, and deeply rooted in, the rest of the living world.

Our first thought on encountering biosemiotic ways of thinking was that perhaps this perspective might offer a general basis for a post-Darwinian theology of nature. But as we looked further into the field of semiotics we began to see some opportunities for rather more specific engagements with Christian theology, and also some interesting consequences for the study of human evolution. From a theological point of view we couldn't help noticing that Peirce's triadic semiotics appears to have some surprising resonances with Christian thinking about the Trinity. To see why, we need to look at Peirce's underlying scheme of metaphysical categories, which he called Firstness, Secondness and Thirdness.

Firstness is the category of sheer being or quality in itself. Examples of Firstness are the sheer quality of a colour, or a taste or smell. Firstness is the category of likeness or resemblance, because likeness is a form of sameness or oneness. It's the category of chanciness and potentiality – the not yet concrete or determined. In terms of the semiotic triad, Firstness corresponds to the character of the object – the as yet unactualized potential for being represented in some way.

Once a sign is set over against the object we have the category of Secondness. This is the category of *distinction* and *otherness*. The sign is *distinct* from the object. Where Firstness is the category of potentiality, Secondness is the category of brute actuality. Things are this way and not that way. The experience of bumping up against reality is an experience of Secondness.

Finally, Thirdness is the category of *mediation*. Where Firstness is the category of potentiality and Secondness is the category of actuality,

[3] Terrence W. Deacon, *The Symbolic Species* (Harmondsworth: Penguin, 2007)

Thirdness is the category of generality and intelligibility. Thirdness is what allows the sign to mediate between the object and the interpreting organism. Without Thirdness the sign could not be meaningful, it could have no significance.

Our initial theological observation, then, is that Peirce's categories of Firstness, Secondness and Thirdness – in more familiar terms, Quality, Otherness and Mediation – appear to have some intriguing parallels with the conceptual structure of Trinitarian thought. The Christian tradition speaks of the second person of the Trinity as the Word of God. As the prologue to John's Gospel puts it: "In the beginning was the Word, and the Word was with God, and the Word was God" (John 1:1-3; NRSV). The eternal distinction of the Son from the Father postulated in Trinitarian thought has its parallel in the irreducible reality of 'otherness' in Peirce's category of Secondness.

Peirce's category of Firstness has its parallel in the concept of the unbegotten or unoriginate nature of the Father, in contrast to the begottenness of the Son. The fourth century Trinitarian controversies revolved largely around the question of whether that which is begotten (i.e., the Son) is necessarily created (and therefore not fully God). From a Peircean perspective Firstness – unoriginateness – need not imply uncreatedness. The world contains Firstness, but it is a created Firstness. And conversely God's being includes *uncreated* Secondness – the Son is 'begotten not made'.

The Church has arguably never quite been sure what to say metaphysically about the Holy Spirit. In John's Gospel Jesus promises that the Father will send the Spirit to the disciples as what is usually translated as an 'advocate' or 'helper'. However, the word *paraclete* can also be translated as 'mediator', which would connect the Spirit with Thirdness.[4] In Acts Chapter 2 the Spirit is the source of interpretation, enabling otherwise unintelligible languages to become intelligible. And the creed we know as the Nicene creed declares the Spirit to be "the Lord, the giver of life," which fits with the biosemiotic idea that it is Thirdness, in the form of a capacity for interpretation, that makes things alive.

There is more that we could say about this 'semiotic model' of the Trinity and how it might be able to do some constructive work in clarifying the perennial conceptual problems of Trinitarian thought.[5] But even then we would have to ask whether it amounts to anything more than

---

[4] William F. Arndt and F. Wilbur Gringrich, *A Greek-English Lexicon of the New Testament and other Early Christian Literature*, 4th revised and augmented edition (Chicago: Chicago University Press, 1952), 623.

[5] See Robinson, *God and the World of Signs*.

a speculative re-working of a piece of rather abstruse theological metaphysics. Does the semiotic model have anything concrete to offer in terms of understanding, say, the significance of the human person Jesus of Nazareth, or what it means to recognise Jesus as the 'Word' of God? This is a challenge that was posed to us by theological colleagues, and it prompted us to investigate whether semiotics might contribute to understanding the Christian concept of the Incarnation.

Semiotics and Incarnation

To explain the way we have developed the 'semiotic model' of the Trinity in an Incarnational direction we need to introduce Peirce's taxonomy of signs. The best known aspect of Peirce's taxonomy is the distinction between icons, indexes, and symbols. This trichotomy concerns the relationship between the sign and object. An icon is a sign that signifies by virtue of a relation of resemblance between the sign and object. For example, a portrait represents a particular person by virtue of a resemblance between the image and its object. An index is a sign that represents by virtue of some direct connection with its object. For example, a weathervane provides a sign of the direction of the wind because the wind causes it to point in that direction. And a symbol is a sign that is assigned to an object by virtue of a rule of interpretation. Language is constructed largely from symbols. If you go to the May Bumps in Cambridge and want to know how your crew is doing you will need to know the rules that connect the colour of the blades to the various colleges. Note that icons are a manifestation of Firstness because they are likenesses; indexes reflect Secondness because they are directly or causally connected to their objects; and symbols are a form of Thirdness because the connection between sign and object is mediated by a rule.

To understand our thinking on the Incarnation we need to introduce a second dimension of Peirce's taxonomy, which concerns the nature of the sign-vehicle *in itself*. A legisign is a sign that is produced according to a rule for the purpose of signifying. The word 'God' is a *symbolic legisign*. It refers to its 'object' by a convention (so it is a symbol), and it is produced according to a rule, by forming the letters G-O-D, so it is a legisign. A knock at the door is also produced according to a rule for the purpose of signifying (tapping one's knuckles on the door – slapping or kicking the door would not be interpreted in the same way) and is therefore a legisign. It is an indexical legisign because there is a direct (causal) connection between the sign (the knock) and the object (the person knocking). The contour lines on an Ordnance Survey map are iconic legisigns. They are produced according to a rule: brown lines of a certain shade and thickness. Other kinds of line would do, but all Ordnance Survey maps follow this convention. However, the lines represent their object – the contours of the landscape – not by arbitrary convention but iconically: they have the same shape as the actual contours of the landscape they represent. So legisigns (sign-vehicles that are produced according to a rule) can have iconic, indexical or symbolic relations to their objects.

Calling on some biblical examples, a celebration of the Eucharist is another, more complex, form of iconic legisign. Each Eucharistic celebration follows a pattern, the 'rules' for which were instituted by Jesus for the purpose of remembering the meal that he shared with his disciples in the Upper Room. The Eucharist is therefore a legisign: it is reproduced according to rule (or set of rules) for the purpose of signifying. However, the relationship between the sign-vehicle (the Eucharist) and object is one of resemblance: each Eucharist iconically resembles the Last Supper or, more generally, the kind of table-fellowship to which those who wish to be Jesus' disciples are called.

Jesus' action in the Temple, gives us an example of an iconic sinsign. A sinsign is a singular occurrence that signifies. Unlike a legisign, therefore, it is not produced according to a rule. Jesus's overturning of the traders' tables was a singular event. He didn't repeat it himself, and didn't suggest to his followers that they should do so. The Temple-action stood for the destruction the Temple. And the relationship between sign and object is again that of a kind of resemblance: the turning over of tables was a minor disruption to the Temple-system that iconically signified a major future catastrophe.

The third kind of way in which something can be a sign-vehicle is as a qualisign. A qualisign signifies solely by virtue of a quality that it embodies. An example is a swatch of cloth acting as a colour-sample for that particular colour. Note that qualisigns can only be icons (not indexes or symbols) because the relation between a qualisign and its object (the quality signified) is always one of resemblance. Similarly a sinsign is always iconic or indexical (not symbolic) because symbols, being related to their objects by a convention, are necessarily more general than singularly occurring instances of signification. On the other hand, as we have seen, legisigns can be iconic, indexical, or symbolic.

The concept of an iconic qualisign is the basis or our proposed approach to the Incarnation. We suggest that to say that Jesus is the "image of the invisible God" (Col. 1:15), the incarnate Word of the Father (cf. John 1:14), is to say that *Jesus's whole life was an iconic qualisign of the saving, transforming presence of Israel's God.* Just as the colour embodied in a piece of cloth may be an iconic qualisign of that very colour, so our proposal is that the person and life of Jesus of Nazareth embodied the very quality of the being of the Father. As an iconic qualisign, the life of Jesus acts as a sign by virtue of being nothing other than the quality that it embodied, namely, the quality of God's transforming presence. Of course Jesus performed *various* signs during his life, as John's Gospel particularly emphasises. Our suggestion is that – taken as a whole – the total 'fabric' of Jesus' life, encompassing all his actions and teaching, all of his "deeds of power, wonders, and signs" (Acts 2:22) constituted an embodiment of the very quality of the Creator's being.

If there is something in this view of the Incarnation as the embodiment of a qualisign of God's being, then it offers a way of connecting the biosemiotic approach to evolution with the heart of Christian theological reflection. The growth of semiotic capabilities in the evolutionary process may be understood as culminating in a particular and distinctive capacity for sign-making and sign-interpretation in humans. And this distinctive semiotic capability of humans may be further understood to have given rise to the event of one particular human, in a particular historical and religious context, having obediently submitted himself to become an ultimate sign of the God whom he called Father. And it is worth noting that for all the complex layers of meaning and interpretation that led up to this event, and which have been involved in its subsequent interpretation, the event itself – considered in terms of the taxonomy of signs – was in some ways the simplest and most fundamental form of signification. That is, it involved merely the embodiment of a quality (a qualisign) that represented iconically (by resemblance) the very quality that it signified. What made this event cosmos-changing was that in this instance this most simple and basic form of representation bridged the ontological gulf between creator and creation: in this unique case the sign-vehicle, the whole fabric of Jesus' life – something firmly on the created side of the gulf – fully and truly embodied the quality of the One who utterly transcends the created order. That, we suggest, is what is meant by Incarnation.

Scientific outworking 1: human evolution

We have touched on some of the potential theological outworkings of the application of semiotic theory to theology. We now turn to consider some scientific applications of semiotics. In this section we begin with a proposal for a new way of thinking about the evolution of human distinctiveness. In the following section we outline our ideas about semiotics and the origin of life.

We have already alluded to Deacon's idea, set out in *The Symbolic Species*, that what makes humans distinctive is our discovery of symbols – the basis of language. But is that the whole story? Consider an image that was much reproduced image in the Darwin anniversary year – a page from one of Darwin's notebooks showing the divergence of species drawn on analogy with the branching of a tree.[6] The diagram clearly shows Darwin beginning to think in evolutionary terms. But note that he is not thinking purely symbolically, but iconically. He is imagining the relationship between species to be *like* a branching tree. We wish to ask, then, what if the thing that makes us distinctive is not merely our capacity to use symbols – which after all is what a bacterium does that when it translates the genetic code – but to combine symbols with other sign types, especially icons? Rather than climbing what we call a semiotic *ladder* – where symbols are thought of as the highest rung, the pinnacle of semiotic competence – we wonder whether human evolution might be better thought of as entering what we call a semiotic 'matrix'.

Let's sketch what this might mean in evolutionary terms. Suppose that our earliest hominid ancestors, such as *Australopithecus afarensis* (around 3-4 million years ago), had learned to communicate by pointing and gesturing. An *afarensis* individual might, for example, have used a recognised hand gesture outlining the shape of a bison's back (an iconic legisign) followed by a pointing gesture (indexical legisign) to indicate the position of an animal being hunted. Perhaps then the earliest species of the human family, *Homo habilis* (from about 2.5 million years ago), may have learned to change some of these *iconic* gestures into *symbolic* gestures, facilitating the speed with which information could be communicated. So instead of outlining a bison shape the gesture might have been reduced to something much simpler – a brief hand-gesture now bearing no direct resemblance to the animal it signifies. Furthermore, gesturing could well have been accompanied by reinforcing vocalisations. And perhaps with *Homo erectus* (from about 2 million years ago) these vocalisations themselves became symbolic, allowing a gradual reduction in reliance on the actual gestures. We might then further suppose that the emergence of

---

[6] Charles Darwin, "Darwin's Notebooks on Transmutation of Species, Part I. First Notebook (July 1837 - February 1838). Edited with and Introduction and Notes by Gavin De Beer," *Bulletin of the British Museum* (Natural History) 2 (1960). Available on-line at http://darwin-online.org.uk/pdf/1960_Notebooks_F1574a.pdf. Accessed 30 March 2010.

fully articulate symbolic speech corresponds to the appearance of anatomically modern humans (around 100 000 years ago).

One of the puzzles of human evolution is why there then seems to be a delay between the appearance of modern humans and evidence of the kinds of cultural advance that we see in the so-called Upper Palaeolithic revolution about 30 000 years ago, whose products include, for example, the beautiful cave art of Lascaux.[7] The story commonly told is that once we had symbolic language we had all the cognitive apparatus necessary for engaging in these 'higher' forms of symbolism. This is the view of human evolution that we call the 'semiotic ladder': a story of the ascent from the use of, say, iconic gestures at the bottom to the peak of semiotic competence, the discovery of how to use arbitrary symbols. According to this view, art, ritual, and so on are examples of the clever things you can do once you are standing on this top rung of the semiotic ladder.[8] But in that case, if full symbolic competence was possessed by the earliest anatomically modern humans, why were the cultural spin-offs of this capacity so delayed?

Our suggestion is that the puzzle of the delay between anatomically modern and behaviourly modern humans dissolves if there was, in fact, a further semiotic threshold to be crossed beyond the discovery of symbolic language. Our idea is that this threshold was the discovery of how to creatively *juxtapose* signs of different types, particularly how to combine symbols with different kinds of icon. Crossing this semiotic threshold – entering the semiotic matrix, as we call it – would have opened up the possibility of ritual, which often involves iconic forms of representation in the context of symbolic competence (recall the example of the Eucharist mentioned above). We are, as Wittgenstein remarked, "the ceremonial animal."[9] And entering the semiotic matrix would also have opened the way to making art – which is the wonderful trick of picking out and representing the qualities, or Firstness, in the world – as we see in the beautiful cave paintings of the Upper Palaeolithic.

---

[7] We acknowledge that recent archaeological finds, such as those in the Blombos cave, suggest that similar developments may have occurred in Africa at earlier dates. Nevertheless, a delay between the first appearance of anatomically modern humans and the flourishing of the kinds of culture associated with the Upper Palaeolithic in Europe remains a reasonable interpretation of the current data.

[8] See, for example, William Noble and Iain Davidson, *Human Evolution, Language and Mind: A Psychological and Archaeological Enquiry* (Cambridge, Cambridge University Press, 1996), 66-67.

[9] Wendy James, *The Ceremonial Animal: A New Portrait of Anthropology* (Oxford: Oxford University Press, 2003), 6-7

The kind of cognitive process we are drawing attention to is not totally unfamiliar – it's the kind of thinking involved in diagrams and metaphors – both of which are kinds of icon that depend on some symbolic competence. What our idea of the semiotic matrix adds is the hypothesis that the distinctive feature of human semiotic competence is not simply that of having learned how to use one of the 'higher' kinds of sign in the matrix (usually understood in terms of the acquisition of symbolic competence), but the discovery of the power of *combining* different sign-types. But what drew us into the semiotic matrix? Our playful hypothesis is that doodling may have played a part. Once our hands were freed from the necessity to gesticulate while communicating we were able absent mindedly to make marks in any available medium while simultaneously speaking or listening. When Christopher Henshilwood reported a 77 000 year old piece of engraved ochre from the Blombos Cave in South Africa, French cave-art expert Jean Clottes dismissed the South African finds as possibly being 'doodles' rather than instances of symbolic behaviour. It is fascinating is that both apparently assumed that the discovery of doodles would be less interesting than the discovery of 'symbolic' art. We would suggest otherwise! Perhaps doodling offered our ancestors the first experience of accidentally juxtaposing symbolic language with externally produced icons – iconic sinsigns – scratchings or scribblings that were produced for no purpose but which took on the function of in some way structuring or shaping our symbolically-mediated thoughts.

Whether there is anything in our doodling hypothesis of human evolution is in fact not the main point of proposing it – the more important point is that it shows how applying Peirce's semiotics to current thinking in palaeoanthropology can, at the very least, clarify some of the conceptual issues involved.

Scientific outworking 2: The origin of life

Another area in which semiotics may be able to clarify a scientific problem from a philosophical perspective is the field of origin of life research. Here we have been able to undertake some empirical testing of our hypotheses. The key idea is that interpretation is one of the properties necessarily possessed by all living things, even the very simplest. In order to show that this idea can actually contribute to the generation of testable scientific hypotheses it has been necessary to show that the concept of interpretation can be given a precise specification that is irreducible to mere effects of mechanistic causation. Elsewhere we have proposed such a definition of interpretation,[10] drawing on T.L. Short's exposition of Peirce's mature theory of signs.[11] According to this account an interpretation is always a *response* of an entity, a response made for a *purpose*. In Peirce's early thinking the response was assumed to be a thought. Peirce later recognised that actions and even feelings could also be interpretants of signs. Thus the simplest interpretative action, we suggest, could be any change of state (such as a change of configuration) in an entity as simple as a single molecule. We propose, therefore, that origin of life research may be advanced by asking what sorts of molecule might be candidates for a 'minimal molecular semiotic entity'.[12]

Our definition of interpretation is irreducible to mere combinations of mechanistic causes because, following Short's account, a purpose is always a 'general' type of outcome. For example, if someone decides to undertake exercise for the purpose of improving their health they may choose a variety of different forms of exercise, any of which may produce the desired general result. Similarly, biological adaptations are genuinely teleological in the sense that a particular structure or behaviour has been selected for a general purpose. Thus the purpose of the heart is to cause circulation of blood. Circulation of blood is a general type of outcome for which, in principle, a number of different particular solutions could have been found. The explanation of the existence of hearts would be incomplete without reference to the causal role played in the evolution of hearts by the general type of outcome of circulation of blood. The heart may be said to exist for the purpose of producing circulation.

Extending this argument, just as some structures 'exist for a purpose', so some changes of state 'occur for a purpose'. Thus a reflex muscle contraction is an adaptation that has evolved for the purpose of, for example, maintaining posture in the context of walking on uneven or unstable ground. An interpretative action is a special case of a change of state that occurs for a purpose. In the interpretative case the response is fallible in the sense that the response is made to some 'sign', X, where X

---

[10] Andrew Robinson and Christopher Southgate, "A General Definition of Interpretation and Its Application to Origin of Life Research," *Biology and Philosophy* 25 (2010): 163-181.

[11] T.L. Short, "The Development of Peirce's Theory of Signs," in *The Cambridge Companion to Peirce*, ed. Cheryl Misak (Cambridge: Cambridge University Press, 2004), 214-240; T.L. Short, *Peirce's Theory of Signs* (Cambridge: Cambridge University Press, 2007)

[12] Andrew Robinson and Christopher Southgate, "Interpretation and the Origin of Life," *Zygon* 45 (2010), 353-358; Robinson and Southgate, "A General Definition of Interpretation," 178.

bears some kind of indirect relation to an 'object', O. An interpretative response may be advantageous compared to a direct response if a responsiveness to X is easier to acquire (evolve) than a direct responsiveness to O. The cost of relying on an indirect form of responsiveness is the risk of being in error. The purpose of the interpretative response will only be achieved if a condition, O, is satisfied, but the responsiveness is to something else, X, which has a fallible relation to O. Hence it might be said that, according to the semiotic view of the origin of life, the property of being capable of making a mistake was a crucial attribute of the first living thing.

We have tested the potential advantage of interpretative responsiveness to a protobiotic entity in computer modelling experiments based on Deacon's concept of an 'autocell'.[13] The autocell is a hypothetical protobiotic entity consisting of two pairs of reactions in each of which two molecules are combined to produce a larger molecule. The system is autocatalytic in that the product of each reaction catalyses the other reaction. In addition, the product of one of the reactions self-assembles to form a hollow shell, which may itself enclose the reactants and product-catalysts. Deacon envisages that periodically the autocell would spontaneously break open, releasing the enclosed molecules and allowing the autocatalytic reactions to resume. We suggested that a mutant autocell that had developed a responsiveness to the presence of substrate molecules in its environment, breaking open on contact with such molecules, could have a selective advantage over the basic autocell. The substrate-sensitive autocell would be interpreting contact with one substrate molecule as a sign of the presence of favourable conditions for 'replication'. In modelling experiments based on hypothetical reaction kinetics for basic and interpretative autocell species we have been able to demonstrate the competitive advantage conferred by the property of interpretative responsiveness.[14] In addition, we have shown the effect of misinterpretation, where autocells break open in response to molecules not associated with the presence of all the necessary substrates for autocatalytic replication.

In another set of experiments our collaborator Niles Lehman at Portland State University, Oregon, has been attempting to synthesise an interpretative ribozyme molecule. Ribozymes are RNA molecules with catalytic activity and form an essential part of the concept of an 'RNA

---

[13] Terrence W. Deacon, "Reciprocal Linkage between Self-Organizing Processes is Sufficient for Self-Reproduction and Evolvability," *Biological Theory* 1 (2006): 136-149.

[14] Leong Ting Lui et al., "Interpretation and the Origin of Life," *Biological Theory* 5 (2010): 112-116.

world' that may have preceded the familiar protein and DNA-based chemistry of contemporary organisms. If RNA chemistry is central to understanding the origin of life then the possibility of an interpretative ribozyme, an RNA molecule capable of making a minimal interpretation of the state of its environment, would be of great interest in exploring the possible role of interpretation in the origin of life. We envisage, for example, that an interpretative ribozyme could exist in two conformational states, 'active' and 'dormant'. In an environment favouring replication it would be advantageous for the ribozyme to switch from the dormant to the active state; an interpretative ribozyme might do this in response to some environmental marker (sign) of favourable conditions. Conversely, in an unfavourable environment the ribozyme would be protected from degradation by remaining in the dormant state. A misinterpretation would occur when the ribozyme switched to the active state in response to a trigger not associated with favourable conditions.

Lehman and colleagues are using the techniques of artificial evolution to select for a ribozyme with these properties, and in preliminary results have already demonstrated both interpretation and misinterpretation.

The semiotic model and natural theology

It is instructive to ask how our use of the philosophical framework of semiotics to engage with scientific and theological problems might fit with the typology of natural theologies suggested by David Fergusson (Chapter 6). It is important to stress, first, that our project is not to be understood as an attempt to construct one of Fergusson's type-1 natural theologies. In other words, we do not claim that, for example, the doctrine of the Trinity could have been deduced from scientific or philosophical efforts independently of God's gracious self-revelation.[15] Trinitarian thought arises primarily from reflection on the significance of the life, death and

[15] As noted in the introductory paragraphs of this paper, one of the original stimuli for this project was a desire to explore the viability of Christian theology whilst taking the implications of Darwinian theology with utmost seriousness. This includes giving due regard to the apparent adequacy of naturalistic explanations of human capacities and their evolution. When we speak here of God's gracious self-revelation it is not necessary to assume that the notion of revelation requires an 'interventionist' understanding of divine action. Indeed, our account of the Incarnation in terms of a capacity of the 'fabric' of the universe to embody the very quality of God's being is formulated in such a way as to allow that God's gift of self-revelation may be conceived of as occurring within the 'natural' processes of the created order.

resurrection of a historical human being, Jesus of Nazareth. Even if one could envisage a scenario in which philosophers had arrived at a Trinitarian understanding of the being of God independently of the economy of salvation as witnessed to in the Judaeo-Christian tradition, there would be no obvious way of deducing the 'content' of the revelation embodied in the incarnation of the Word in this person Jesus. Another way of putting this is to say that even if it were proposed on philosophical grounds that God's eternal Word might become embodied in the fabric of creation, still there would be no way of deducing by means of reason alone the 'quality' that would be thereby revealed. Only the actual quality embodied in the whole life-death-resurrection of the incarnate Word could reveal the quality of God's being.

Neither should our project be regarded as a type-2 natural theology, since we do not seek to construct a scientific or philosophical framework against which all theological claims must be evaluated. In other words, we are not seeking to subordinate theology to science or philosophy. Nevertheless, to the extent that certain theological and scientific / philosophical frameworks are found to cohere, we believe that the latter (scientific / philosophical) may give some measure of support to the former (theological) without being their principal ground of legitimation.

Moving to the other end of Fergusson's typology, our framework is clearly not simply a (type-4) rebuttal of the claims of the New Atheists, though it challenges some of the core assumptions of the new atheism. For example, a demonstration that general types (rather than only particulars) can be causally efficacious undermines the assumption that all of the world's processes can be understood entirely in terms of mechanistic causation. In other words, the semiotic framework and its scientific outworkings support the view that the emergence and effects of purposes are a real and irreducible aspect of the natural world. Of course the fact that purposes emerge 'immanently' in the world does not guarantee that their ultimate origin lies in the purposes of a transcendent creator. But it does refute one possible line of argument against the existence of a purposeful creator, namely that apparently teleological aspects of the world are in fact illusory.

The semiotic approach is, similarly, not merely an example of a type-5 natural theology, though we certainly seek to demonstrate the potential compatibility of certain theological, scientific and philosophical lines of thought. Importantly, though, we seek to press the semiotic approach beyond the question of passive coherence and to move onto the task of positively stimulating new work in both theology and science. In this we are indebted to the model of 'creative mutual interaction' developed by

Robert J. Russell at the Centre for Theology and the Natural Sciences in Berkeley, California.[16] Thus on the theological side we have attempted to develop a new approach to Trinitarian thought based on Peirce's categories and semiotics, and a new way of understanding the Incarnation drawing on Peirce's taxonomy of signs. On the scientific side we have used the same philosophical resources as the basis of a proposal for a new approach to the problem of the origin of life and as a new way of understanding the nature of human distinctiveness and the mechanisms of human evolution.

The idea of 'creative mutual interaction' brings us finally to Fergusson's type-3 approach to natural theology. We are much taken with Sarah Coakley's call (Chapter 7) to regard Fergusson's type-3 natural theology as an opportunity to challenge what she calls the 'flat-plane' assumption of the other approaches; that is, the sense in which the epistemological playing field has been assumed to have been marked out in advance by either the scientific or theological side (usually the former). Coakley suggests that:

> ... this [Type-3] ontological destabilization of the flat plane dogma needs to be complemented – in my view – by a different, and sophisticated, epistemological and hermeneutical one, which really takes on working scientists where they are, and, as a matter of apologetics, challenges them to explain the metaphysical and narrative presumptions on which they themselves operate. (p. 99)

We suggest that our project fits well with this kind of approach. We employ Peirce's semiotics and metaphysics as a framework from which to question the metaphysical presuppositions behind diverse areas of scientific enquiry. In the field of origin of life research we ask scientists to consider the problem in terms of the question of the simplest conceivable interpretative entity. In the field of human evolution we call into question assumptions about the 'highest' form of semiotic competence. We do so from within an epistemological perspective informed by reflection on core elements of Christian theology in the light of the (religiously satisfying) hypothesis that meaning, purpose and signification are ontologically primary elements of the fabric of the universe. We do not claim (epistemologically) to be able to deduce the content of God's gracious self-revelation from purely scientific or philosophical premises. Rather,

---

[16] See, for example, Ted Peters and Nathan Hallanger, eds., *God's Action in Nature's World: Essays in Honour of Robert John Russell* (Farnham: Ashgate, 2006).

we regard the (epistemological) basis of the possibility of knowledge of the world and of God to be the (ontological) structuring of the fabric of the world as a reflection and vestige of the being of the transcendent creator. This is a bold hypothesis, certainly; a hypothesis that, to borrow Coakley's words in this volume, does more than "embrace 'modest', optional, interpretative glosses of a theological kind on normatively atheistical science" (p. 100). We hope that in doing so it does offer, again to quote Coakley, "a more convincing and creative alternative for Christianity today" (p. 100).

The semiotic model and glory

One test of whether our semiotic approach will ultimately prove to be convincing is whether it turns out to be capable of illuminating areas of scientific and theological enquiry beyond those that have comprised our initial areas of investigation. As an illustration of how such extensions of our approach might be pursued we conclude with some brief exploratory remarks on the nature of 'glory'.

In his paper in *Zygon* in 2004, and more fully in his book *God and the World of Signs*, published after the present paper was originally presented, Robinson has developed the idea that Peirce's categories of Firstness, Secondness and Thirdness might allow a re-expression of the notion of the Father creating through the two hands of the Son and the Spirit.[17] But it is far harder to see how this thinking could do work in respect of God's healing of creation, and the drawing of humans, and all creation, into participation in the divine life, such that God becomes in the words of 1 Corinthians 15 "all in all". Southgate has begun to explore how such questions might be addressed, particularly in terms of how we might understand the encounter with Firstness. This thinking follows an interest in the biblical concept of 'glory'.

We're accustomed of course to think of glory in terms of very bright light. But fascinatingly the underlying Hebrew concept, and in particular the word *kavod* which is the dominant term for glory in the Hebrew Bible, has much more the connotation of 'weight', and hence of importance, of depth of reality, unignorable reality. Much more the sense, in fact, of gravitas than of lightness of being. *The* great study of glory in the modern theological corpus is of course Hans Urs von Balthasar's theological aesthetics – an extraordinary and somewhat unmanageable ocean of

[17] Andrew Robinson, "Continuity, Naturalism and Contingency," 127; Robinson, *God and the World of Signs*, 248-254.

insight. In looking at the great theophanies of the Hebrew Bible from Sinai onwards Balthasar notes that they are not first characterised by divine *speech*. The primary impact on Moses at the burning bush, on Elijah at Horeb, on Isaiah in the Temple, is one of sheer presence, evoking tremendous awe. In the terms of Peirce's categories this understanding of glory might be regarded as a manifestation of Firstness, of encounter with the sheer quality of the entity. From there, it might be said, God recognises, gives attention to, the presence of the creature in its otherness (Secondness), its limitedness and fallibility. But this gracious recognition, allowing the creature ways to continue to live in the face of the utter holiness and awesomeness of the divine, takes place – has to take place – through some mediating relation: in Peircean terms some manifestation of Thirdness. Classically in the Hebrew Bible this is through the giving of the Law and the operation of cultic worship.

Encounter with Firstness, apprehension of the sheer quality of entities, may also be possible through intense contemplation of the natural world. This is what Gerard Manley Hopkins attempts to characterise in his language of 'inscape' and 'instress'. The inscape of an entity includes what it is scientifically, its existence, form and pattern, but also its particularity, its 'thisness' (a term Hopkins particularly emphasised). All of these are part of its createdness. The contemplative thinker has the capacity to 'instress' an entity by giving – handing over – his/her attention completely to the entity. So Hopkins can write "I do not think I have ever seen anything more beautiful than the bluebell I have been looking at. I know the beauty of our Lord by it."[18] The sheer quality (Firstness) of the bluebell makes its createdness transparent, and through it the divine Word, the source of otherness (Secondness) in creation, is glimpsed. This is made possible by the work of the Spirit, who leads the explorer into all truth (Thirdness).

In Paul we see a rather different emphasis. Now the focus is that in having the mind of Christ, being hidden with Christ, knowing Christ the hope of glory, the believer is transformed from one degree of glory to another. A very important insight of Balthasar's is that "there is no seeing without being caught up"[19] – our contemplation, in the fullest sense of the word, of Christ as qualisign of God in his self-giving, self-emptying life of love, catches us up into that life. To contemplate Christ and his Cross

---

[18] Gerard Manley Hopkins, *Poems and Prose of Gerard Manley Hopkins*, ed. W.H. Gardner (Harmondsworth: Penguin, 1953), 122.

[19] Hans Urs von Balthasar, *The Glory of the Lord: A Theological Aesthetics Vol VII: Theology – The New Covenant* (San Francisco: Ignatius Press, 1989, German original 1969), 24.

teaches us the full stretch of the love by which God engages with the creature, and it brings us into participation in what Michael Gorman terms the 'cruciformity' of the divine. The character of God, God's glory, proves in the Christian vision to be that glory of self-surrender to the other that we call the life of the Trinity. And that self-surrender within God makes space for the creature to live in freedom.

There may be a helpful analogy to be drawn with listening to a piece of music. To begin to listen to, say, a Beethoven string quartet is to enter a world full of relationships and structuring rules, full of Secondness and Thirdness. But those make possible an encounter with Firstness, with the sheer quality of what the music expresses, a quality irreducible to its component elements. The believer then is drawn into the space of conversation that is the quartet, so intimately that as T.S. Eliot put it, in a sequence itself inspired by late Beethoven quartets, in "the unattended moment" music is "heard so deeply / That it is not heard at all, but you are the music / While the music lasts."[20]

Thus also by giving over one's attention not – to continue the quotation from Eliot – to the flux of "usual pastimes and drugs," but to the life of Jesus in its fullness, the believer finds him/herself in the midst of a triune conversation full of histories and relationships, but ultimately and irreducibly expressible as a quality of love.

This sort of formulation should not be allowed to seem to privilege contemplation over action. The contemplation of the face of Christ through worship is a form of relating in freedom to the God who made us and saves us; it is itself transformative and takes us deeper into the glory of that God. But insofar as anyone sees profound need and responds to it, reaches out to the poor, the sick, the hungry or imprisoned, that action reveals someone already caught up into the divine conversation, already acting in the image of the God of self-giving love. Other-regard and worshipful contemplation, then, are two sides of the same coin.

One last observation: in the area of salvation through the contemplation of Christ we are dealing with a very special sort of sign, a sign that both reveals and transforms (clearly this has parallels with what sacramental theology has classically tried to say). Whereas the theophany of the burning bush casts Moses to the ground, and the beauty of the bluebell moves Hopkins to an apprehension of the Lord as Creator, Jesus' life is the sign that stretches out its arms and draws the seer in. This is the limit case of semiosis as it flowers into theosis – sign, object, interpretant and

---

[20] T.S. Eliot, *Collected Poems and Plays* (London: Faber and Faber, 1969), 190.

purpose collapse into one another. This indeed is the Christian understanding of the work of the economic Trinity.

Conclusion: vestiges of the Trinity?

To sum up, if we put together our semiotic model of the Trinity with the biosemiotic approach to evolution we think that quite a rich theology of nature begins to emerge. There is a highly suggestive strand in the Judaeo-Christian tradition which understands creation as not merely a work of God the Father but as also mediated by the Word and the Spirit. Irenaeus talked about the Word and Spirit being the two 'hands' of God in creation, and Augustine developed the idea that it might be possible to discern 'vestiges of the Trinity' in creation. We suggest that the processes of semiosis and interpretation in the world may be understood as such vestiges of the Trinity in creation. According to our proposal the long and costly process of biological evolution may be thought of as a growth in the richness of interpretive processes. The origin of life may be understood as the emergence of a capacity of proto-biotic entities to make a minimal form of interpretation of the state of their environment. Human distinctiveness, we suggest, is deeply rooted in this semiotic impulse of nature. And the structure of the semiotic processes that the world supports is such that entry into the semiotic matrix opens a window onto that which transcends the created order. According to this view the nature of the world is exactly the structure required for the emergence of meaning; the emergence and subsequent development of life depends upon an increasing capacity for meaning-making and meaning-seeking. Theologically, God is thus the ground of all meaning and the ultimate goal of the universe's emerging capacity to interpret signs

A crucial juncture in the history of evolution was the point at which the fabric of the world had evolved to a point at which a particular creature was able to function as a *qualisign* of the love of the creator, and at which fellow creatures were – and are – able to make appropriate interpretive responses to that sign:

> In the beginning was the Word, and the Word was with God, and the Word was God. All things came into being through him, and without him not one thing came into being. ... And the Word became flesh and lived among us, and we have seen his glory, the glory as of a father's only son, full of grace and truth (John 1: 1-3, 14; NRSV).

CHAPTER TEN

RESPONSE: THE METAPHYSICS OF CO-OPERATIVE ENQUIRY

KENNETH WILSON

Building on insights drawn from the work of C.S. Peirce, Andrew Robinson and Christopher Southgate suggest that science and religion are not only in permanent dialogue, but interdependent and mutually supportive. Peirce's ultimate purpose in life was to do all that he could to enhance human understanding. For him, the key was open enquiry co-operatively engaged in, employing all the dimensions of human experience. He explicitly rejected Descartes' attempt to ground secure knowledge in what he could not doubt – namely his own existence – because that assumed that private experience was paramount. His own work witnesses to his embrace of co-operative, ranging as it does from logic (which he regarded as a normative science), to empirical investigation, metaphysics, ethics, and not excluding theology.

Continuity and co-operation

Peirce denied discontinuities, and by affirming continuities assumed an ultimate if ungraspable coherence in all enquiry. Mere rationalism was therefore an inadequate methodology, since its assumptions did not allow for synechism – the coherence of all explanation in principle into one systematic account – or tychism, Peirce's belief that no single account could be ultimately true because the universe was open to chance events. Hence while deduction is a valid method of enquiry – in principle it does no more than confirm what we already know – and induction was already open to question in the light of experience, our basic task was always what Peirce called abduction. He meant by this the 'musing' on the way the world must be if it was to give rise to human experience of the kind we enjoy. It was this 'musing', he believed, that gave rise to a (rather unorthodox) appreciation of God's existence, and a triadic view of the nature of reality existing in abstract qualities, existent things, and their relationship. He grasped after this via semiotics, which he claimed allowed human beings to be aware of the true nature of things through signs. The human being is distinctive precisely in having the capacity to interpret signs. In this whole process, Peirce claims, we are not restricted to immediate discoveries: co-operative enquiry must include being open to the whole history of human enquiry if we are to gain purchase on reality. The primary influence on Peirce was Kant, though Thomas Reid was important too, as were Plato, Aristotle, St Augustine, Aquinas, Ockham and Duns Scotus.

The approach of Robinson and Southgate, in the light of their appreciation of Peirce's complex position, bears witness to Michael Oakeshott's view of education as "a conversation between the generations."[1] As a result there are a huge number of ways in which this paper could encourage conversation to develop. Without, therefore, any intention of frustrating the responses which the paper will have stimulated in others, I shall focus on what I take to be three crucial themes.

First, it supports the view that theology is a vital dimension of human enquiry; secondly, it links the ways whereby there emerged the human capacity / desire to give itself an identity by means of the processes of evolution; thirdly, it develops the creative insight and consequent value which flows from employing the particular notion of interpretation implicit in the metaphysics of Charles Peirce to ground and illuminate Christian theological perspectives and provide a framework for understanding the human condition.

The rhythm of question and answer

So, first, theology is an essential dimension of human enquiry. Theology is no longer regarded as 'the Queen of the Sciences'. Indeed there are those who would argue that sensible human beings have grown out of it. After all, what interest can there be in systems of thought based on unverifiable claims about the world supposedly revealed directly by God which turn out, in so far as they are intelligible at all, to be in direct conflict with the publicly testable claims of scientific enquiry? Of course,

[1] Cf. Michael Oakeshott, "Education: The Engagement and Its Frustration," in *The Voice of Liberal Learning: Michael Oakeshott on Education*, ed. Timothy Fuller (New Haven and London, Yale University Press, 1989), 76-80.

such an understanding of the role of the doctrine of revelation in the Christian tradition is quite mistaken. In fact, as Keith Ward has argued, the doctrine of revelation opens enquiry, it does not bring it to an end. It is the underpinning perception that the truth of whatever it is that we think we know depends upon the outcomes of further enquiry, ad infinitum – or maybe we should say ad deum: there is more to discover. Peirce would have approved! Moreover, successful enquiry is the outcome of careful attention to the evidence, the discernment of the relationships inherent in the business of learning, and the determination to see things through responsibly. As Bernard Lonergan puts it in his marvelous book *Insight*, our understanding depends upon being attentive, being intelligent, and being responsible.[2] Seneca was quite right when he wrote, "We must learn as long as we lack knowledge, which means as long as we live." This is not inconsistent with the role of enquiry as Peirce understood it.

On the basis of this paper we may say that the experience of being human is not defined by an inherent capacity to receive indubitable truths from the Almighty but by the growing awareness that we are capable of intelligent enquiry, the outcome of which is the ability to interpret the signs we intuit and thereby to explore responsibly the space in which we find ourselves, physical, material, moral and spiritual. The overall thesis advanced by Robinson and Southgate is that:

> the nature of the world is exactly the structure required for the emergence of meaning; the emergence and subsequent development of life depends upon an increasing capacity for meaning-making and meaning-seeking. Theologically, God is thus the ground of all meaning and the ultimate goal of the universe's emerging capacity to interpret signs. (p. 144)

Moreover, it is argued, the thesis is empirically testable through research in palaeoanthropology.

There are valuable comparisons, I believe, with R.G. Collingwood, who is receiving more attention currently than for some time. Collingwood, in *Speculum Mentis*,[3] offers an account of the map of human knowledge which he suggests can be traced both in the history of the race and in the life of the individual. Collingwood speaks of "the rhythm of question and answer as constituting the very life of the spirit." Human knowledge is the product of the process by which humankind (and an individual person) becomes self-aware; that is, spiritually self-aware. It

---

[2] Bernard Lonergan, *Insight: A Study of Human Understanding* (New York: Philosophical Library, 3rd edition 1970).

[3] R.G. Collingwood, *Speculum Mentis* (Oxford, Oxford University Press, 1924).

begins with the speculative creation of works of art which he thinks is the cutting edge of the mind, the perpetual reaching of thought into the unknown – I was struck by Robinson and Southgate's speculative reference to the possible role of doodling in human evolution. Works of art are infinite in number – they lead to the emergence of religion as the celebration of the reality of the ultimate object of enquiry, namely God. This proceeds through science (experience ordered into an enquiring focus) and on to historical enquiry, which is the attempt to place all knowledge in a time sequence. This in turn gives rise to questions about purpose and the morality of behaviour in the context of the sustaining and constantly re-telling of the human narrative, and then ultimately to philosophy and, I would add, theology, which includes in principle all possible knowledge and self-consciousness, involving both where I am, who I am, and what I think.

None of these aspects of human enquiry is sufficient by itself since we are concerned as human beings to locate ourselves, to discover ourselves in the environment in which we have been put. For this to be achieved we require to use manifold co-ordinates. A crucial theme of this intriguing paper is the claim that we shall be unable to locate ourselves, know ourselves, until we have learned to relate to The Other – that which is over and beyond us, which is God, Father, Son, and Holy Spirit. The space in which we live and move and have our being is one in which God is the living presence which gives it its character. God is not an outsider entering into our world, but the Creator whose grace makes the world how it is, and makes it possible for us also to know how the world is and thus begin to fulfil the purpose of God in creating. God, as Trinity, is wholly other (the Father), wholly embodied (the Son), and wholly present (the Holy Spirit). Moreover, given its basis in Peirce's semiotics and his concept of 'musing', the account which Robinson and Southgate offer of both Trinity and Incarnation is both illuminating and revelatory.

The evolution of the human self

Secondly, there is the linking of the emergence of human self-consciousness with the natural processes of evolution. We become aware of ourselves, our world, our relationships with it and are able to recognize, question and interpret the signs we pick up and explore them in art, religion, science, history and philosophy. And we are able to make responsible judgments about it all because our evolution makes us aware of just how close our relationship is with the world of which we are a part. What is more, there is nothing accidental about this.

Of course, the whole world might have been different, in the sense that there might not have been self-consciousness embodied in the particular shape and of the particular kind that we know as human being; but it is inconceivable that there would not have emerged a self-conscious search for meaning in some form or another because that is how the world is. The linking of the emergence of meaning-making and meaning seeking to the evolutionary process is of profound significance. Its essential importance lies in the fact that it demonstrates just how the development of cognition and the ability to 'recognise' and interpret signs takes place progressively over the whole period of human evolution. Our growing self-awareness is the product of a growing capacity to interpret signs. This makes possible 'experimental living': living which provokes questions, proposes solutions, builds habits of behaviour in the light of experiment, gives rise to ways of thinking about ourselves which are of value, and gives human being a critical advantage for survival. These ideas seem consistent with those developed, for example, by Nicholas Humphrey in his recently published *Seeing Red*. He says, "Consciousness matters because it is its function to matter. It has been designed to create in human beings a Self whose life is worth pursuing."[4]

Lord May, theoretical ecologist, formally President of the Royal Society, reflected recently that religious belief may have been profoundly important in human evolution as a means of controlling our aggression towards the environment and to one another – he even suggested that we might benefit from another dose of it.[5] Andrew Newberg is exploring the biology of religious experience while Michael Persinger of Laurentian University, Ontario, used powerful magnetic fields to produce visions and spiritual experiences among volunteers.[6] This might suggest that religious belief is hard-wired into us and of possible benefit to human survival: it does not of course show that God exists. But Robinson and Southgate show, in contrast with Humphrey, Lord May, et al., that religious belief and theological enquiry matter not just because it has turned out to be of

---

[4] Nicholas Humphrey, *Seeing Red* (New Haven and London: Belknap Press at Harvard University Press, 2006), 131.

[5] Lord May made this remark in a speech on the eve of the annual meeting of the Royal Society in 2009. It was made with regard to his sense of the calamitous trajectory on which the world was set because of the failure of the world to agree and implement means of coping with climate change.

[6] Cf. Andrew Newberg, *Why God Won't Go Away* (Ballantine Press, 2001); Andrew Newberg, *Principles of Neurotheology* (Burlington, VT. and Aldershot: Ashgate, 2010); Michael Persinger, *Neuropsychological Bases of God Beliefs* (Westport, Conn., Praeger, 1987).

advantage to us in the evolutionary pattern of things but because the evolutionary process is itself embodied in a world which is graciously made by God so as to create the possibility of this outcome.

Trinitarian theology and Peircean metaphysics

Thirdly, there is the grounding of the interpretative value of Christian theology, with particular reference to the Incarnation and the Trinity, in Peirce's metaphysics. The claim is that this not only provides an illuminating insight into the productive use of theological enquiry but, via semiotics, makes real connections with the everyday world of our experience as explained in evolution. Robinson and Southgate outline the three categories which underpin Peirce's semiotics. 'Firstness', the quality itself – e.g. colour; 'Secondness', the otherness of the object, e.g. egg-cup; 'Thirdness', which links quality and object, e.g. indicates generality. Each is independent but all inherently one because of the unity of all knowledge of the one world. Herein lies a fruitful way of understanding the Trinity. The Father is the essential quality of life, the Son is the Other which embodies the very quality of the being of God, and the Holy Spirit is the generality which links the two, and enlivens all things. No separation here, but no simple identity either: rather a coherent eternal relationship.

The semiotics developed by Peirce suggests a promising way of – I don't quite know how to put it – understanding, explaining (?), describing, or identifying the space in which claims of Incarnation make sense. Signification involves a triadic relation of object (Firstness); sign (Secondness); interpretant (Thirdness). Jesus, in Peircean terms, is not a legisign – a symbol for God, or a sinsign – a resemblance or an icon of God in some sense; Jesus is an iconic qualisign; that is, he in his entire life is nothing other than the embodiment of the whole quality / character of God. Moreover, since the quality which characterizes God is dynamic, redemptive and salvific, the freedom which Jesus embodies instantiates the encouraging character of God's creation in his incarnation. He therefore reveals the nature of the divine, the character of God's creation and God's living eternal, affectionate presence – in fact the Trinity.

The space which we inhabit is transformed by the consciousness that the quality of the divine is not independent of the human, but embodied in Jesus whom we call the Messiah. One might put it differently by saying that in the decision to create, God made room for the 'other', the creation, and by so doing enabled the other to make room for God, which is what we mean by the Incarnation.

All this underlines the active role which human enquiry and human behaviour plays in God's process of creating: our influence may be good if it is with the grain and contrary to it if we choose to ignore the true dimensions of our environment, which includes God – conceived of and experienced as Trinity. This thesis is liberating as well as true!

The fact – I presume the claim is that it is a fact – that Jesus in his entire life is the quality of the divine – means that we can by the habits of our behaviour, including of course our intent to understand more fully our place in the world of God's creating, embody in ourselves and in the world the same divine quality. Or does it? Are we talking of the divinization of the human? Athanasius affirmed that God became man in order that man may become divine. Certainly in the Eucharist we not only celebrate the fact of God's presence with God's people, we also receive the body and blood of Christ. There is a Real Presence.

The more I reflect on what Southgate and Robinson have produced in this paper, the more disappointed I am that Peirce, who wrote much in his life but published little, never completed a full statement of his philosophy. There's gold in them there hills.

Part III:

Further Dialogues with Darwin

CHAPTER ELEVEN

BARTH AND DARWIN: CAN THEY TALK?

PHILIP CHAPMAN

This paper proposes that it is possible is to bring into conversation with each other two understandings of the human person: that of Karl Barth and that of Darwinian biology. The problems with this exercise centre on the difficulty of finding a starting point for a discussion between two enterprises which appear to have nothing in common and which cannot in advance be assumed to acknowledge each others' intellectual respectability. The hope of success is that they are both intending to answer the same question: What is humanity? An obstacle is that science can pursue its course without attempting to understand the person as self or subject, though most scientists would regard this as desirable. Barth takes a realist view of the human subject but relates it much more closely to its physicality than might have been expected.

What methodology should we adopt? An obvious strategy would be to look for a philosophical framework for the conversation. However, Barth completely denies the possibility of any philosophical basis to theology. The subject matter of theology is quite simply the disclosure of the divine by the divine, but more than this and indispensably, the revelation of God in Jesus Christ as mediated by Holy Scripture. Further, 'nature', however understood, has nothing to say to theology, unless, and this is a vital qualification, seen within the overall Christocentric apprehension of creation. For Barth, the purpose of the observed and observable universe is to be the theatre for the performance of God's saving action in Jesus Christ. Barth divides the created cosmos into two spheres. The upper sphere, 'heaven', is inaccessible to human enquiry. The lower sphere is the one to which to which the scientific enterprise now known as 'cosmology', as well as all the other sciences, relate. This is the sphere of the natural. According to Barth, humanity lives at the boundary between earth and heaven, the natural and that which is beyond nature.

Secular parables of the truth

In Kantian fashion, Barth regards the sciences as limited to the study of phenomena.[1] On this basis his revelatory theology has ontological and epistemological priority: science would be the junior partner in any dialogue, but one who must be listened to. Revelation can offer us particular flashes of insight concerning the truth of God and God's dealings with humanity, though we attempt to systematize them at our peril. Such revelatory truth tells us nothing directly about scientific matters: these are the province of human rationality. However, there is no such thing as an independent rationality, or indeed any other human capacity, apart from the accompanying acts of the same Spirit who instils the soul's life. The Spirit does not guarantee scientific truth, but enables the possibility of setting the phenomena in order.

God is, however, sufficiently generous with truth. Whether or not it is aware of it, humanity will have what it needs. Moreover the discernment of truth is not restricted to a small class of Christian believers: the operation of the Spirit in the revelation of truth is available for the whole human race. There can even be "secular parables of the truth,"[2] and secular prophets. These latter may well be atheistic practitioners of the sciences who alert the Church to realities she may have failed to discern. Here we are still talking about theological truth. Yet in so far as that truth is about the human person, whose existence does indeed have a physical component, the exploration of phenomena, be they physical, chemical, biological, psychological or sociological will, whether recognized or not, be in the environment of the transcendent truth of God.

For these reasons, the rationality which patiently explores and attempts to understand the order in the phenomena of the world can be understood as Spirit-enabled.[3] In fact humanity is summoned to this task of ordering.[4]

---

[1] Karl Barth, *Church Dogmatics*, translated and edited by G.W. Bromiley, T.F. Torrance and others (Edinburgh: T&T Clark, 1937-1975). Hereafter cited as CD followed by Volume and Part. This reference is to CD III(2), 121-2.

[2] See George Hunsinger, *How to read Karl Barth* (New York and Oxford: Oxford University Press 1991), 234-280, referring to CD IV(3.1), 38-165; see also CD IV(3.1), 235. For secular parables of the truth especially in relation to other religions see Paul S. Chung, "Karl Barth and God in Creation: Towards an Interfaith Dialogue with Science and Religion," *Theology and Science* 3 (2005): 56-69.

[3] According to Hunsinger Barth's thought closely parallels that of Calvin. Where Calvin speaks of "the Spirit of God" in this context Barth speaks of the prophetic work of Jesus Christ. Calvin talks about "the human mind, Barth about human words. Hunsinger, *How to Read Karl Barth*, 234

This would entitle the theologian to agree with the scientist that there is a common task, which each explores in their own fashion: the task of making sense of things. From the scientist's point of view the search for better understanding of the world as it presents itself to our senses may be its own justification. Alternatively, science may be the means toward some practical or humanitarian goal. Yet, from the Barthian point of view, making sense of the phenomena must be vital to us: this is the world we inhabit, the only one we have. It is in this realm that we have to listen for and obey the command of God. On Barth's premises the more we know of what is going on, as far beneath the immediate surface of things as we can get, the fuller our obedience can be, whether or not we are believers. However, we are not entitled to treat the natural world as being purely for human benefit. There has to be a proper respect for the works of the Creator.

Such a programme of research must be distinguished from later interpretation of the Baconian project, beyond the Puritan context in which it first took root, and from related apologia for the imperialist diversion of nature to human ends: Barth does not assume that the world is there simply for exploitation and enjoyment by humanity. On the contrary, each created entity has its own relationship with the Creator, one that is not transparent to us, and we must assume that everything has its own particular way of being. Further, the command of God enjoins respect, even reverence for life. We should, Barth says, approach the slaughter house with fear and trembling.[5] Scientific endeavour, when subjected to ethical constraints, is a proper activity, but a purely human one. Consequently such truth as is displayed by it is *ipso facto* relative but can, under grace, point beyond itself. It follows that one important task of theology is to attempt to discern the secular parables of the truth in the scientific accounts of things.

Science and theology compared

There is thus in Barthian terms a justification of the scientific enterprise if properly conducted. Science, is one set of probes that humanity uses to make sense of things, and where possible, to put that understanding to use. There is no single method or procedure. There is no

---

[4] CD IV (3.1), p. 147

[5] CD III(4), 324-356. Barth here acknowledges a debt to Albert Schweitzer. For the way in which Bacon's successors have gone far beyond his intentions see Neil Messer, *Selfish Genes and Christian Ethics* (London: SCM, 2007), 224-225, with references adduced there.

one pattern by which hypotheses are elaborated and tested. There is no unique criterion of success. But research is constrained by its social and economic context as well as by political and personal pressures. More fundamentally, and this is especially obvious in the case of quantum physics, the observer and what is observed are unavoidably connected. There is, even in theory, no way of observing without affecting what is to be observed. Ironically even those who wish to deny the reality of the human subject – relegating it to the realm of so called 'folk psychology' or replacing it by talk of the physical states of neurones – cannot remove themselves from the subject's point of view, any more than physics can.[6] Even non-theological or anti-theological science cannot escape from its essential humanity. All these factors reinforce Barth's view that the findings of scientific inquiry are relative: they are a result of human activities. They cannot, of themselves, be a revelation of truth. Barth can live with even the most thoroughgoing naturalistic explanations since they are incapable of disclosing how things are in themselves.

Theology is also a human enterprise. That its data are obtained from revelation, as Barth would assert, does not make it infallible. But nonetheless, from its own point of view theology has a priority over other kinds of knowledge since its primary subject matter is the self-disclosure of the living God. Consequently it does deal, not in phenomena, but in how things are in themselves, and in particular with the reality of God and God's action in the lower cosmos. As an academic subject it proceeds, like others, by discussion and publication, and suffers largely from the same contextual constraints. Formally speaking, in this account of Christian theology, examination of the witness to revelation recorded in Holy Scripture fulfils an equivalent function to the checking and repetition of investigations in empirical science. Materially the validation of theological assertions is by faith in the self-revealing God, and consequently in God's revelation through the Word. Indeed, theological thinking is a form of prayer. Barth would not deny that here there is a hermeneutical circle involved; such knowledge is only attainable, or rather receivable, by an assumption of its possibility – Barth makes no apology for this. Even though human knowledge as such is restricted to the phenomenal, the noumenal can indeed be known: at this point Barth breaks fundamentally with Kant.

---

[6] e.g. Paul M. Churchland, "Eliminative Materialism and the Propositional Attitudes," in *Mind and Cognition*, ed. William G. Lycan (Malden, Mass. and Oxford: Oxford University Press, 1999), 120-132

Consequences

If, as intended, our methodology follows from Barth's standpoint, it must be aware that faith needs no justification from outside itself. This is not, therefore, an exercise in apologetics. Faith does, however, have an obligation to communicate itself and to do so rationally. But there are limitations both for science and theology. For Barth there is no way in which the whole of created reality in its two spheres can be viewed within a common epistemological framework. We cannot therefore fulfil Stephen Hawking's ironically pious wish to "know the mind of God."[7] Further, although from outside faith itself we might think of faith's assertions as formally similar to the way in which mathematical argument proceeds, namely axiomatically from premises to conclusion, for Barth this would be a mistake. There can be no procedure for knowing God. What humanity in reality is can only be known through the revelation in the person of Jesus Christ. Barth's fundamental statement about the human person is that "Man is the soul of his body," Jesus Christ being the prototype human. Barth understands both the soul and the physicality with which it is inextricably intertwined as essential to the reality of humans in their finitude. Barth regards the body as, in some sense, the form of the soul: indeed the body has a certain priority over the soul. This soul is not a substance: it is what the Spirit does in the human body. Barth has not rejected the examination of human phenomena, merely the assertion that the truth about the human person can be arrived at in this way. In fact not only naturalistic, but also philosophical and other interpretations of humanity must be read as "commentary on a text which must first be known."[8] They may be correct as far as they go but only God, the Creator, can tell humans what they are: God must speak and we must listen to his Word.

So what kind of commentary, then, might science, in the succession of Charles Darwin, make on an understanding of humanity which must be primarily theological? Can 'nature' be allowed, and even encouraged, to speak to theology within the Barthian framework which is often understood as hostile to natural theology? Such speech is possible. Indeed, we can bring critically appraised scientific accounts into relation with basically theological ones. Such scientific accounts need not be true in themselves in any absolute sense, but they can point to truth. A given scientific statement might function as a secular parable of truth in the way

---

[7] Stephen W. Hawking, *A Brief History of Time – from Big Bangs to Black Holes* (London: Bantam Press, 1988), 175.

[8] CD III(2), 122

already mentioned. We can, for instance, see the possibility that Charles Darwin was in Barth's sense a secular prophet. If the theologian quotes Genesis 2:11, "Then the Lord God formed man from the dust of the ground," then Darwin's prophetic commentary on this might be:

> In the distant future I see open fields for far more important researches. Psychology will be based on a new foundation, that of the necessary acquirement of each mental power and capacity by gradation. Light will be thrown on the origin of man and his history.[9]

For the theologian a lesson, not just from Darwin but also from his successors in evolutionary biology and neuroscience, as well as in cognitive psychology, is that the fundamental physicality of the human person must be taken with utter seriousness. Humanity cannot be in some fashion above this. Human persons, though ensouled as Barth holds, are, in a reality to which Darwin points, finite and close kin with the rest of created nature: Barth and Darwin (and the Darwinists) can talk. It is a conversation that needs to be further developed.

---

[9] Charles Darwin, *The Origin of Species*, ed. J.W. Burrows (Harmondsworth: Penguin, 1985), 458.

CHAPTER TWELVE

TEILHARD DE CHARDIN: FROM NATURE TO SUPERNATURE

MICHAEL BURDETT

Pierre Teilhard de Chardin, the French palaeontologist and Jesuit, was a dynamic thinker and person. At once inspirational and convivial, he offered an invigorating incorporation of evolution into Christian theology during a time of scepticism towards both religion and human progress. While his approach to evolution is more influenced by and reminiscent of Bergson than Darwin,[1] he still remains one of the most important people to have considered evolution and its import to Christianity to date.

This paper tracks the general thrust of Teilhard's programme, which begins with naturalistic presuppositions, but culminates in a theology of nature. In particular, I want to draw attention to the way in which Teilhard's book *The Phenomenon of Man* begins methodologically with a strictly scientific perspective. In utilising this scientific methodology Teilhard asserts that a strictly naturalistic account of the world gives way to a more spiritual perspective. In a sense, one might say that science culminates in the spiritual and the inward. From this movement, it becomes apparent that something like a theology of nature is the expected and demanded perspective that the Teilhardian must invoke when considering the natural world.

My primary focus will be on this transition in the text and how it then gets worked out in Teilhard's eschatology. Finally, several comments will be made which address the extent to which Teilhard can be used in our current discussion of Darwin and a theology of nature, as well as some criticisms Teilhard might have for us today in this discussion.

---

[1] See Henri Bergson, *Creative Evolution*, trans. Arthur Mitchell (New York: Henry Holt, 1911).

Teilhard's '*within*' and '*without*'

Very early on in *The Phenomenon of Man* Teilhard draws a distinction between two aspects of the world. Both of these realms have corresponding histories which are related to one another. The first is related to what Teilhard refers to as the *without*. The *without* is simply that which is external and physical in the world. The *without* is the world of matter and is governed by scientific and natural laws. Chemical laws, physics and biology are the prime way of understanding the *without*. Indeed, Teilhard's position as a palaeontologist and scientist provides the framework for this aspect of the world. The *without* is the scientific aspect of Teilhard's dialectic. Teilhard's account of cosmological and terrestrial history relies heavily upon this more scientific aspect of this dialectic. For instance, in talking about the creation of the planets and the development of life on Earth Teilhard does not refer to biblical texts, but to current scientific theory. His methodology in arriving at man is proper to the history of the *without*.[2]

Teilhard's basic phenomenology not only includes that which is empirically observable, but is inclusive of the other end of this dialectic, which is not so empirically obvious. In fact, just as Teilhard's vocation as a scientist lays the groundwork for his interpretation of the *without*, perhaps his role as a Jesuit informs this other end of the dialectic. The opposing side is termed the *within* of things. As implied, the *within* is not something one can observe with one's eyes, ears, or other bodily senses, but is nevertheless just as important and real. The *within* is an aspect of all reality. Indeed, all material contains a *within* and is inherent in everything. Just as all physical matter has a particular mass, density and volume so too does all matter contain a *within*. The *within* is somewhat more convoluted and strange when applied to the more simplistic forms of matter. For this reason Teilhard suggests that one should first consider the phenomenon of man so as to better grasp why the *within* is a necessary aspect of the whole of reality.[3]

When one considers humans in relation to other species it becomes apparent that a significant capacity distinguishes us from other animals. The *within* is directly related to this new capacity. This new capacity is

---

[2] For a more detailed explanation of the without see, in particular, Teilhard's first chapter entitled "The Stuff of the Universe" in Pierre Teilhard de Chardin, *The Phenomenon of Man*, trans. Bernard Wall (New York: Perennial Library, 1975).

[3] See Teilhard's chapter "The Within of Things" in *The Phenomenon of Man*; also Emile Rideau, *Teilhard De Chardin: A Guide to His Thought* (London: Collins, 1967), 88-92.

merely the most complex instantiation of the *within* in history: human consciousness. And with the discovery of this new phenomenon comes the need to make room for a new category in the universe. As Teilhard suggests, "consciousness is evident in man … therefore, half-seen in this one flash of light, it has a cosmic extension, and as such is surrounded by an aura of indefinite spatial and temporal extensions."[4] Human consciousness does not fit into the general naturalistic view of the world when taken from the perspective of the *without* alone. Something more is implied in human consciousness. Teilhard therefore posits a new category of reality. In short, human consciousness implies a *within*.

The advent of human consciousness is not only important to our awareness of a distinct realm in the world – the *within* – but something dramatic occurs in human history with its arrival. Indeed, it is the appearance of thought which necessitates a categorical interruption of a strictly naturalistic view of the world. As has been noted, this interruption is not something which only takes place with the appearance of man, but is even present amongst more primordial forms of matter where it acts as a forerunner to the emergence of thought and consciousness.[5] As Teilhard puts it, "The principle I mean is that consciousness is the peculiar and specific property of *organized states* of matter; it is a property that cannot be detected, and may therefore be neglected in practice, with low values of complexity, but that gradually emerges and finally becomes dominant, with high values."[6] According to Teilhard, then, with the advent of consciousness something is brought to bear in the world which represents more than a mere interruption in the advancement of a contiguous naturalistic and physical history. An entirely new and qualitatively different Earth is created with the arrival of consciousness.

The noosphere and the future of humanity

The spontaneous emergence of thought does not signify merely a local phenomenon on the face of the earth. As Teilhard claims, humanity "is more than a branch, more even than the kingdom; he is nothing less than a 'sphere' – the noosphere (or thinking sphere) superimposed on, and co-extensive with (but in so many ways more close-knit and homogeneous)

---

[4] Teilhard de Chardin, *The Phenomenon of Man*, 56.

[5] Teilhard claims this is most evident in behaviour and instinct found in the animal kingdom. See *The Phenomenon of Man*, 164-168.

[6] Pierre Teilhard de Chardin, *The Future of Man*, trans. Norman Denny (New York: Image Books/Doubleday, 2004), 167.

the biosphere."[7] As humans spread across the world their unique faculties are seen to pervade it as well. So much so, that it affects more than just the area of humanity – it institutes "a transformation affecting the state of the entire planet."[8] It is for these reasons that Teilhard chooses to use the word 'noosphere' when referring to this phenomenon. The noosphere is the collective mind of humanity as thought emerges from the human species. The concept of the noosphere concerns the phenomenon of psychic energy from the planetary level. The noosphere is the thinking and conscious domain of collective humanity.

What is important for our purposes, beyond just the recognition of a new area in reality, is that the *within* and the noosphere become the most efficacious factors in determining the future of the human species. Indeed, according to Teilhard, the *without* and the trends associated with it primarily governed the shaping of history prior to humanity, but with the arrival of humans the *within* becomes the most efficacious force. Teilhard contends that unity and convergence are mechanisms of the *within* while divergence and separation characterise the *without*. Or, to put it historically, prior to humanity evolution followed a divergent trend following a fan-like structure, but with humanity and our faculty of consciousness the trend then shifts towards convergence. Teilhard claims, "a convergent mankind is identical with an ultra-reflective mankind: and, conversely, an ultra-reflective mankind is identical with a convergent mankind."[9] Teilhard asserts that it is the development of the noosphere which sparks such a radical change. As human thought at the terrestrial level becomes more advanced and complex with education, technology and socialization, so understanding the world becomes less dependent upon strictly biological explanations. If the depth of humanity's past reveals a world governed by DNA, evolution and biology it is the noosphere and the spiritual which govern its present and future.

Christology, creation and convergence

The sphere of the *within* has been introduced above. I now turn to consider its religious overtones. Indeed, Teilhard's spirituality is observed most predominantly in this sphere, as reflected in works such as *Hymn of the Universe* and *Le Milieu Divin*. The history and future of humanity does

---

[7] Pierre Teilhard de Chardin, *Man's Place in Nature: The Human Zoological Group*, trans. René Hague (London: Collins, 1966), 80.
[8] Teilhard de Chardin, *The Phenomenon of Man*, 181.
[9] Pierre Teilhard de Chardin, *Activation of Energy*, trans. René Hague (London: Collins, 1970), 327.

lie within the purview of science. But, more importantly, the history and future of humankind is not complete without considering the religious aspect of history and the future – for this is just as, if not more, efficacious in sparking the movement to the eschaton and the divinisation of nature. This religious aspect of nature finds its culmination and origin in the coming of Christ to the world and has massive repercussions for the natural world.

The historical Christ's appearance on earth enacts a very real change in the ontology of nature. This change is rooted in the very hypostasis of the person of Christ. For, as Christ breaks into the world his divinity ruptures the enclosure of nature and creates a spark which causes a chain reaction. His divinity mingles with the natural and becomes immanent within it. The Eucharist is not only the reflection upon this coupling of divinity and matter, but in a very real sense is the prototype for the divinisation of matter. As Teilhard exclaims, "through your own incarnation, my God, all matter is henceforth incarnate."[10] In this way, through the advent of Christ the supernatural becomes an aspect not only of humanity's history, but also of matter and even biological forces.[11] This seriously changes the course of natural history. The religious becomes important for the scientist because of Christ's incarnation and the flood of the divine into the world.

This Christological propagation in the world is related directly to Teilhard's notion of creation. Rather than expressing a static occurrence at the beginning of history, Teilhard regards creation as a process enacted through the mystical Christ as it transforms the world into greater divinity. Evolution is the biological mechanism by which God is in process of creating the world and moving deeper within it. Teilhard uses the term Christogenesis to signify the advancement of the mystical Christ in history, which not only parallels biological evolution but, because of Christ's hypostasis, *is also* biological evolution. Teilhard has been accused of emphasising theosis and the creative function of Christ over his redemptive work on the cross, but for Teilhard these cannot be separated from one another. As he says,

> Christ, it is true, is still he who bears the sins of the world; moral evil is in some mysterious way paid for by suffering. But, even more essentially, Christ is he who structurally in himself, and for all of us, overcomes the resistance to unification offered by the multiple, resistance to the rise of

[10] Pierre Teilhard de Chardin, *Hymn of the Universe*, trans. Simon Bartholomew (London: Collins, 1965), 24.
[11] See Rideau, *Teilhard de Chardin*, 163.

> spirit inherent in matter. Christ is he who bears the burden constructionally inevitable, of every sort of creation.[12]

The redemption of humanity from sin is seen through the lens of Christ's creative and transformative work in nature more generally. Precisely because God has chosen to create this world, through Christ, he also redeems it from the cosmic sin of counter-evolution and the lack of evolutionary progress.[13]

According to Teilhard this spiritual future of humanity also rests upon the phenomenon of convergence. In Teilhard's eschatology the noosphere becomes increasingly personalised and centred until, at a point in the future, humanity will reach an apex which he refers to as the Omega Point.[14] This is identified with the cosmic Christ of Pauline theology.[15] This entirely religious future reveals the spiritual dimension of our world here and now as it expresses the *telos* and, by extension, the essence of history and nature. It is from this perspective that Teilhard develops his more mystical and religious works, such as *Hymn of the Universe* and *Le Milieu Divin,* which rely upon a theology of nature. These works never lose the perspective that the world around us is divinized and contains a *within* – indeed, if one follows Teilhard through this transition, the natural world cannot be considered any other way.

Teilhard today

It remains to ask how far Teilhard can be utilised today? I have already mentioned that Teilhard's understanding of evolution follows Bergson rather than Darwin. While Bergsonian evolution might have been a more viable interpretation of evolution at the time of his writing, it is its Darwinian counterpart that has been generally accepted by the scientific community. Bergson's evolution invoked extra-material mechanisms, in particular the *élan vital*, to account for certain phenomena in evolution, claiming Darwin's model could not account for it merely through mutation and survival of the fittest.[16] However, it is this creative force – God and

---

[12] Pierre Teilhard de Chardin, *Christianity and Evolution*, trans. René Hague (London: Collins, 1971), 85.
[13] For a further account of Teilhard's interpretation of original, moral and cosmic sin see Rideau, *Teilhard de Chardin*, 169-172.
[14] Teilhard de Chardin, *The Phenomenon of Man*, 257-272.
[15] Ibid., 293-294; Pierre Teilhard de Chardin, *Le Milieu Divin: An Essay on the Interior Life*, trans. Bernard Wall (London: Collins, 1960), 105.
[16] See Bergson, *Creative Evolution*, 87-97.

the spiritual – that scientists might have issue with today. Furthermore, Teilhard's view of ontogenesis and phylogenesis – the idea of an evolutionary progression of all species as a whole – would also be scientifically suspect. Most scientists would not claim that certain species are more advanced than others or that evolution shows progressive directional tendencies. So, while it might seem on the face of things interesting to invoke Teilhard in the science and religion discourse, unfortunately his scheme will more than likely be rejected.

Furthermore, scientists may take issue with Teilhard's attribution of religious and spiritual significance to consciousness and the internal life. Does consciousness really imply anything religious? It seems that this move is crucial in Teilhard's work. It is much easier to attribute spiritual significance to the *within* when one comes from a historical tradition which associates the inward life with religion and the spiritual. But what if one does not concede that consciousness is inherently religious or spiritual? Much seems to depend on one's initial presuppositions. It is clear that a Teilhardian theology of nature does not follow as a logically necessary outworking of a naturalistic, scientific, starting point. Nevertheless, one need not necessarily be deterred from following Teilhard in his religious approach to nature, superimposing such an approach on the scientific account.

I suggest that what is convincing about Teilhard's approach, even if it does rely upon an outdated understanding of evolution, is that a purely biological approach to evolution is no longer sufficient to describe the phenomenon of humankind. The biological perspective might explain aspects of our past, but cannot do justice to our future. This can be partially explained by recognising that humanity's prime environment is increasingly becoming less and less biological. Indeed, to draw for a moment on Jacques Ellul's sociology, the natural environment gives way to the social environment, which in turn becomes the technological environment.[17] Teilhard, in his more eschatological works, contends that it is precisely the social and technological nature of humanity that is the prime locus in determining humanity's future.

Finally, then, how might Teilhard's approach offer a critique of modern evolutionary science? Might one say that Darwinian evolution cannot really adequately explain the phenomenon of humankind and its future in purely biological terms? That is, indeed, the position for which I am arguing. For even if biological evolution is accepted as the guiding

---

[17] Jacques Ellul, *What I Believe*, trans. Geoffrey William Bromiley (London: Eerdmans, 1989), 89-140.

principle for natural history, I suggest that it becomes increasingly inadequate when focusing on humanity. For example, it seems that the biotechnology and information technologies are able radically to transform humanity today much more quickly than natural evolution ever could. Indeed, humanity can spark its own evolutionary development through bio-enhancement and gene therapy. Furthermore, information technologies are transforming the very social nature of humanity, so much so that describing the social nature in purely biological terms is becoming a thing of the past. I suggest, therefore – in line with Teilhard's final thought on the natural world – that invoking a purely naturalistic and biological account of ourselves is insufficient for understanding the world and our future and leaves room for something like a theology of nature to be developed.

PART IV:

DARWIN'S LEGACY: OPPORTUNITIES AND CHALLENGES

CHAPTER THIRTEEN

DARWIN'S SEVEN GIFTS TO THEOLOGY

FABIEN REVOL

In this paper I suggest that current evolutionary biology presents at least seven gifts to theology, enabling theologians to enter into peaceful constructive dialogue with Darwin's scientific heritage. These gifts are:

1. A new general framework within which to think about nature.
2. The end of static conceptions of the universe, allowing us to envisage nature more dynamically and historically.
3. The possibility of thinking about the design of nature at another level, i.e. by looking for it in the general tendencies of a universe which becomes continuously organised rather than in the details of particular natural structures.
4. The eviction of non-Christian images of God which follows from the failure of the design argument.
5. The recognition of God's humility and self-emptying as a response to the problem of theodicy: the sign of the cross as the answer to the question of ontological evil.
6. The possibility of a new concept of finality, understood as an immanent desire for form.
7. The recovery of the theme of immanence in theology, allowing us better to appreciate the continuous and active presence of God.

I suggest that these seven gifts are the signs of the passage of natural theology to theology of nature.[1] That is, these gifts invite a move towards interpretation of the theological meaning of the history of nature and away from attempts to discover or deduce the existence of God and his attributes from features of the natural order.

---

[1] The conference at which the papers in this volume were presented had as its subtitle "From natural theology to a theology of nature."

Gift 1: A new framework for thinking about nature

With Arthur Peacocke,[2] I notice some undeniable data that theology has to take into account in its dialogue with biology: First, that evolution is not merely a theory but a fact; second, it is a fact that is to be set against a broader panorama, which presents the cosmic evolution of the entire universe. Third, that the development of applied mathematics in the sciences of nature by means of statistics allows an understanding of how chance and necessity complement each other in allowing nature to be creative of novelty and complexity. Fourth, that Darwin's work contributed to showing the continuity between humans and animals, a continuity that has been subsequently confirmed, for example, in the fields of molecular biology and ethology. Fifth, that the importance of the concepts of reduction and complexity in science are now acknowledged: every level of organization owns its proper complexity and laws. Sixth, that there is nowadays an awareness of ecology as the science of integration of living systems in organized wholes.

A consequence of these developments is that biological evolution is not exhausted by the "mutation/selection" model, but rather includes also an ethologic approach. As a result, evolution provides a new framework for thinking about nature which emphasises the intrinsically dynamic character of the created order. This leads us to the second gift.

Gift 2: The end of static conceptions of the universe

The new view of nature that follows from the evolutionary perspective invites the theologian to leave behind a view of the cosmos as an arena of static order. Instead, evolution shows a world in permanent construction, not yet finished, wherein order and disorder unceasingly mix. The cosmos understood as a finite and ordered whole shatters into pieces in front of the reality of evolution. The paradigm which best describes the world is not the one of order according to the Greek philosophers, but of contingency,[3] with the emphasis shifting to the notion of probabilistic order. To illustrate this idea, Dominican friar Jacques Arnould says that the theologian was chased away from the earthly paradise, which represents the theme of a statically ordered cosmos, to begin a journey towards meeting the challenges posed by the sciences.

---

[2] Arthur Peacocke, *Evolution, the Disguised Friend of Faith?: Selected Essays* (Philadelphia & London: Templeton Foundation Press, 2004), 36-42.

[3] Ibid., 278-279.

Gift 3: Thinking about design at another level

Connected to this notion of cosmic order we find that of design. In a pre-Darwinian conception of the world, order refers to organization that in its turn refers to an organizer. This is the approach to natural theology well known in Britain through William Paley's *Natural Theology*.[4] Charles Darwin's contribution points towards a world certainly not ordered according to the usual sense of design, but regarded, on the contrary, as a vast construction site of do-it-yourself. There is here a paradoxical third gift: Darwin's objections to design in nature actually raises the question of the level of the design. If the design of nature really followed Paley's notion of design then we should expect to find a perfect functioning nature, flawless, without waste, without imperfections such as pain, death, parasitism, etc. This is the notion of design that Darwin rejects.

Ian Barbour picks up the question of design in nature and offers an interesting solution. Arguing that the order of nature corresponds to a footprint of the divine intelligence in creation, he thinks that design is not to be found in the details of nature but in general tendencies.[5] He suggests, in other words, that some general laws are given, and that the development of creation can then be open to true novelties through trial and error. For instance, the improvement of the living through evolution is not determined by its perfect and idealistic form. Its perfection consists in going through the wayward adjustments between the general order and the mechanisms of the chaos, thus participating in the invention of its own form. In this perspective, chance and necessity are complementary aspects of the design of the cosmos. Barbour thus identifies a continuing pattern of creation which "is a trial-and-error experiment, always building on what is already there."[6]

Gift 4: The eviction of non-Christian images of God

Two images of God fall with the Darwinian challenge to the argument from design. First, the God of the gaps: if there is no design in nature of the kind envisaged by Paley then there is little place for the notion of a god who works through direct intervention among the natural causes of the

[4] William Paley, *Natural Theology, or Evidence of the Existence and Attributes of the Deity, collected from the appearances of nature*, eds. Matthew D. Eddy & David Knight (Oxford: Oxford University Press, 2006).
[5] Ian G. Barbour, *Religion and science: Historical and Contemporary Issues* (New York: HarperSanFrancisco, 1997), 238.
[6] Ibid., 240.

world. This gives real autonomy to the scientific and theological discourses. In fact it means that God's activity within creation is not to be understood in the terms of scientific explanation and discourse. For the scientist, it means that God's activity is not concurrent with or complementary to natural laws; it even escapes any scientific inquiry. For theologians it means also that the defects of sciences are not to be explained by theology, but rather that science can present questions to faith as it interprets its own traditions of looking at the world.

The other non-Christian image of God which is called into question by the Darwinian perspective is that of a judge who imposes his laws of nature as he does the laws of human morality. The design argument implies an understanding of God's creative work in terms of legislative activity, understanding the laws of nature as reflecting a divine legislative decree. As Paul Santmire and John Cobb put it:

> Just as God laid down the laws according to which both earthly and heavenly bodies move, also God laid down the laws for human behavior. The difference was that material entities necessarily obey the laws of God, whereas human beings are free to obey or to disobey. However, whereas the laws of earthly rulers are sometimes broken with impunity, the laws of the divine rulers must always be enforced in the end. Hence, it was evident to many that at death all would be judged according to their obedience or disobedience and rewarded or punished. This belief was understood to be both rational in itself and important for the maintenance of good behavior and social order.[7]

In other words, the idea of divinely decreed laws of nature, and the associated idea that the details of creaturely form and behaviour reflect the intentions and design of God, may be used to support an understanding of God as one who will judge humans according to their degree of conformity to divinely instituted religious laws. Calling into question the argument of design may therefore help Christians to leave behind this non-Christian representation of the God-judge, and to therefore focus on the God of Jesus Christ, about whom Scriptures state that He is essentially love and mercy.

We must acknowledge, however, that the Darwinian perspective, by bringing to attention the apparent waste and suffering involved in the evolutionary process, may also be held to undermine the idea of a loving God. This brings us to the fifth gift, which concerns the issue of theodicy.

---

[7] H. Paul Santmire and John B. Cobb Jr, "The world of nature according to protestant tradition," in *The Oxford Handbook of Religion and Ecology*, ed. Roger S. Gottlieb (Oxford: Oxford University Press, 2006), 125.

Gift 5: The importance of God's humility and self-emptying

The image of a good God who takes care of his creatures seems to explode with Darwinism. Darwin himself could not conceive that a good God could have produced the plans of a world in which so much suffering and useless death could take place.[8] The image of a God of love cannot be found in nature. This is a major challenge for theologians.[9] Nevertheless, this point can be considered as a fifth gift. Indeed, American Catholic theologian John Haught argues that nature has a cruciform configuration; that is, it has the form of the cross of Jesus. Haught calls for development of the biblical theme of God's humility. For Haught, God is a compassionate God, affected by the suffering of his creatures. Recalling Schillebeeckx, he says that "God's humility does not imply weakness and powerlessness, but rather a kind of 'defencelessness' or 'vulnerability'."[10] This humility theme attains its full sense in the context of the hymn of the epistle to the Philippians (2:6-11): Kenosis means this renunciation of a dominating power, capable of embracing the struggles of the whole universe not only of the human race. Associated with this assumption of suffering is the hope of the promise of its redemption accomplished in the resurrection.

Gift 6: The possibility of a new concept of finality

According to the concept of natural selection there is no finality, direction, or final causation in the mechanisms of the evolution. This may appear to imply that God does not care for creation, that he does not steer its progress. Such an understanding would empty the concept of providence of its meaning. This annoys theologians, who generally hold

---

[8] It must be noted that this image of violence in nature has been widely replaced now thanks to the recent developments of ecology. This scientific discipline emphasises that predation was a natural relationship among many others of collaboration, mutualism and symbiosis, showing thus the importance of positive interactions in nature over mere violence. See for example Gary L. Miller and Robert E. Ricklefs, *Écologie*, trans. Michel Baguette et al (Bruxelles: De Boeck, 2005), 11.

[9] Ted Peters and Martinez J. Hewlett, *Evolution from Creation to New Creation: Conflict, Conversation, and Convergence* (Nashville: Abingdon Press, 2003), 22-24.

[10] John Haught, *God after Darwin, A Theology of Evolution*, Second Edition (Boulder: Westview Press, 2008), 52.

that one cannot conceive that there could be creation without a purpose for the creation. In that case it may be necessary to admit that the ultimate end or purpose of nature is not perceptible within nature itself. For example, Ted Peters, an American Lutheran theologian, suggests an eschatological viewpoint according to which the ultimate meaning of the creation and its end are given in its relationship with God.[11]

An alternative theological response is to argue that nature does, in fact, exhibit a kind of end-directedness. In spite of the impact of the works of Darwin, contemporary sciences may be understood as showing that there is an internal dynamism in nature, a dynamism that is directed towards the production of novelty. This novelty is contingent because it depends on the interaction between the laws of nature and the chance intersections of the elements of a causal series. Emmanuel Gabellieri argues that the concept of nature shifts from an essentialist, fixist connotation, to a dynamic connotation, in which processes are structured in ways envisaged in the process philosophy of Whitehead and the philosophy of the *élan vital* of Henri Bergson.[12] The concept of nature that blossoms from this approach recalls that of Aristotle's *fusij*, in which movement and rest come together; that is to say, it suggests that nature has in itself a principle of development towards its perfection. At the metaphysical level, it is possible to consider that this end obeys a final cause. According to Bertrand Souchard there is an end – that he considers as a relative end – in nature;[13] and the orientation towards the end is the passage, for every creature, from potency into act:

> Nature is a principle hosted in the thing itself. By potency and act, of movement and of the living, it is the progressive (and immanent) actualization of beings which Aristotle notices. This internal end means that matter is appetite of form, a propensity to actualization.[14]

---

[11] Peters and Hewlett, *Evolution from Creation to New Creation*, 27-31.

[12] Emmanuel Gabellieri, "Nature et création: Opposition ou nouvelle alliance ?", in *Nature et création entre science et théologie*, ed. J.M. Exbrayat and E. Gabellieri (Paris: Librairie Philosophique J.Vrin-IIEE, 2006), 181-230; cf. Alfred N. Whitehead, *Process and Reality, an Essay in Cosmology*, Corrected Edition, eds. David Ray Griffin and Donald W. Sherburne (New York, London: The Free Press [1929] 1978); Henri Bergson, *Creative Evolution*, trans. Arthur Mitchell (New York: Henry Holt, 1911).

[13] Bertrand Souchard, *Aristote de la physique à la métaphysique, réceptivité et causalité* (Dijon: Éditions Universitaires de Dijon, 2003), my translation.

[14] Ibid., 254.

It seems to me that in this expression, "matter is appetite of form," we have the essential summary of what Aristotle means by finality. Matter is appetite of form, but we cannot necessarily say which one a priori. Only the contingent history of matter and of life can tell it. The relative character of finality corresponds indeed to the integration of the notion of chance as properly belonging to the way of functioning of natural phenomena. It is in this case possible to rescue the concept of finality from its determinist yoke. We can then understand that in nature the end is a priori opened upon a plurality of possible forms. And this open quest for new forms is manifested in the different evolutionary ways with their successes and failures.

Now, for Aristotle (in the *Metaphysics*) God is the pure act, and he draws to himself all beings which are in actuation. God, as pure act, is also final cause. And God as final cause would act on nature according to an attraction that would cause beings to pass from potency into act, according to the mode of desire. Even if chance introduces an element of indeterminacy into natural phenomena, the desire connected to the final cause is effective, bringing creation out of passivity to make it enter into an evolutionary dynamics, and so giving it finality, the quest for form. This may be the meaning of evolution: it is an answer of creation to God's appeal to it to continue its process of self-accomplishment.

This question of finality brings us to the question of God's action in creation.

Gift 7: The return of the theme of immanence

The Tractarians of the Oxford Movement welcomed the theory of evolution as a blessing because it allowed them to promote a vision of God as present in all creation and whose action becomes intermingled with the secondary causes of nature. This approach, called 'immanentist', is opposed to a purely transcendent view of God where God is understood to be outside creation. For the Tractarians, the Darwinian approach was a blessing in that it allowed them to articulate God's presence and action in creation more fully than the notion of the special creation of species would allow. While the evolutionary view implied the continuity of God's presence, the idea of special acts of creation implied long phases of absence:

> The scientific evidence in favour of evolution, as a theory is infinitely more Christian than a theory of "special creation". For it implies the immanence of God in nature, and the omnipresence of His creative power. Those who oppose the doctrine of evolution in defence of a "continued

> intervention" of God, seem to have failed to notice that a theory of occasional intervention implies as its correlative a theory of ordinary absence.[15]

Compared to the consequences of an over-emphasis on divine transcendence, this immanentist approach seems more consistent with the vision of a God of love who wants to be close to his creatures.

Conclusion

Darwin's seven gifts to theology combine to form an invitation to pass from natural theology to theology of nature: that is, to seek to interpret the theological meaning of nature's history rather than attempting to discover or deduce the existence of God and his attributes from the causal details of nature.[16] Receiving those gifts for our time, it is now possible fully to acknowledge Aubrey Moore's words as Arthur Peackoke did in one of his later books:[17]

> Science has pushed the deist's God further and further away, and at the moment when it seemed as if He would be thrust out altogether, Darwinism appeared, and, under the disguise of a foe, did the work of a friend.[18]

[15] Aubrey L. Moore, *Science and the Faith: Essays on Apologetic Subjects, With an Introduction.* 1893. A facsimile of the fourth edition (London: Kegan Paul, Trench, Trübner & Co.), 132, 151-152.
[16] Jeffrey P. Schloss, "Evolutionary theory and religious belief," in *The Oxford Handbook of Religion and Science*, eds. Philip Clayton and Zachary Simpson (Oxford: Oxford University Press, 2006), 200.
[17] Peacocke, *Evolution, the Disguised Friend of Faith*, 32-34.
[18] Aubrey L. Moore, "The Christian Doctrine of God," in *Lux Mundi.* 15th edition, ed. C. Gore (London: Murray, 1904), 73.

CHAPTER FOURTEEN

OVERVIEW: NATURAL THEOLOGY AND THEOLOGICAL NATURALISM

NEIL SPURWAY

Darwin's legacy

Our symposium opens, as it had to do, with Charles Darwin himself, figured in the chapter by 'Sam' Berry and in Celia Deane-Drummond's responding comments. The mild and tentative genius, who has so influenced humanity's conception of itself, is persuasively depicted in Sam's early pages.[1] So are both the brilliance of his insight and its inevitable incompleteness in a world which had no understanding of genetics. Sam does service also in stressing that Mendel's remarkable work did nothing to help, when first it became known. We can see now that one of the reasons for this is that the phenomena Mendel studied – distinct traits in the mature plants, dependent on single genes – though essential to initiate the new science, were seriously atypical representatives of the gene-phenotype relationship. Hence many, though not all, of the scientific vicissitudes of Darwinism in its first 70-80 years.

The social vicissitudes, the political and religious reactions, were inevitably more diverse than the scientific ones. It is worth noticing, however, that the antiquity of the earth, assumed and required by Darwin, was rarely an issue even for his most vociferous contemporary critics: modern Creationists are more anti-scientific in this respect than the majority of their Victorian predecessors. As David Fergusson notes in his later chapter, by 1859 the pioneer geologists – Hutton, Lyell, Cuvier – had

---

[1] I hope I shall not cause offence by departing from normal practice and referring to contributors by their first names. As all but two are personal friends, I find stark surnames inappropriate.

convinced their many voracious readers about the reality of deep time. This concept was consonant, too, with the relatively deep space (though actually still confined to our galaxy) revealed by astronomers such as Herschel. Even the scriptural geologists, committed to reading the record of the rocks in terms compatible with Genesis, had by the 1850s largely adopted the 'day-age' theory, which interpreted the days of the first chapter of Genesis as each referring to a geological aeon. Hugh Miller's *Testimony of the Rocks* (1857) went so far as to match them up:[2] Day 3, for instance, when "the earth brought forth grass, the herb yielding seed, … and the tree yielding fruit," represented for Miller the Carboniferous when the trees flourished which would later form our coal-layers, while Day 5 referred to the Oolitic and Cretaceous, when great reptiles and primitive birds dominated the habitable land and the air immediately above it. In fact, it was not scripture but physics which provided the main challenge to the timescale implied by *The Origin,* for Lord Kelvin, considering the Earth to be a fragment detached from the Sun, but ignorant of the radioactivity which is its internal source of heat, calculated that it could not possibly have stayed warm enough to support life for anywhere near as long as Darwin's idea required.

Closely related to the acceptance that the earth is very old was the concept of evolution *per se*. This had been clearly expressed at least 60 years before *The Origin* – not least by Darwin's grandfather Erasmus, but also by Lamarck (and anticipated by other French thinkers some 40 years before him). Charles Darwin's original contribution was to propose a *mechanism* for evolution very different from Lamarck's. The mechanism was the occurrence of variations between progeny and parents which was random, not dependent on the behaviour of those parents (this was the distinction from Lamarck), but would be inherited by the generations arising from those progeny; and the action upon the variant forms of straightforward competition for all the necessities of life: food, living space and mates. This competition gave rise to 'natural selection' – those better adapted to their circumstances would, on average, produce more offspring. The process was fundamental and inescapable: *natural selection cannot not occur!* The questions were and are whether it is the only mechanism leading to differences between generations, and whether it is powerful enough by itself to produce the changes which have occurred. I shall return to these questions at the end of this piece.

---

[2] Hugh Miller, "The Testimony of the Rocks; Or, Geology in its Bearing on the Two Theologies, Natural and Revealed," reprinted in *Creationism and Scriptural Geology, 1814-1857*, ed. John M. Lynch (Bristol: Thoemmes Press, 2002)

Imago dei

Meanwhile, fun though it is to supplement Sam Berry's story, my chief task is to highlight, and comment upon, the main points he himself makes. Like Celia Dean-Drummond, I read Sam as finding the main challenge of Darwinian thought to a religious world-view as lying not so much in the process of creation as in its account of the nature of humanity – in its matching of biological to theological anthropology. Sam writes from a conservative evangelical standpoint, which is distinctly different from Celia's and even further from my own. Nevertheless, it is important to recognize that he finds little difficulty in reconciling such a stance with (neo-)Darwinism, and that there will probably be many readers who derive real comfort from this fact. In taking any issue with him, therefore, I want to be seen only as exploring an alternative stance intellectually, not as demeaning the spiritual satisfaction which Sam's position gives to him, and to those who share his cast of mind.

That said, I pick out for comment the parenthesis in Sam's allusion to "a purely naturalistic – which tends to imply atheistic – origin for human kind" (p. 30). Evangelicals less broadly informed than Sam, and probably all fundamentalists, accept that implication uncritically, whereas for me a naturalistic account of any aspect of being human is, quite simply, the only sort of account which can be correct. But to say that it is correct is not to say that it is easy to formulate. Consciousness is extremely hard to understand naturalistically, and it is only a little less difficult to present an entirely naturalistic account of morality or of the broader, non-sexual and non-familial, aspects of love. Yet close observers of higher primates – most recently, in public awareness, Frans de Waal [3] – describe many instances of such love, and of the beginnings of ethical behaviour. Every owner of a horse or dog is quite clear that their animals love them. And consciousness, though never provable, seems to most of us to be present in a wide range of living creatures far less advanced than chimps, horses or dogs. For me, the *imago dei* simply is the sum of such properties: the sum is larger in a human being than in any other animal, but the difference is one of degree not kind. The properties summated are all natural, all have arisen with infinite gradualness in the course of evolution, and none has involved a special providential intervention at a specific point in time. The *whole evolutionary process* is God-given, not any individual step.

---

[3] Frans M.B de Waal, *Good Natured: The Origins of Right and Wrong in Humans and Other Animals* (Cambridge, MA: Harvard University Press, 1996); Stephen Macedo and Josiah Ober, eds., *Primates and Philosophers: How Morality Evolved* (Princeton, NJ: Princeton University Press, 2006).

So, when Sam says that "God's image is a divine attribute. It is not an evolved character. Although its expression depends on our bodies, it is not genetic," I cannot go along with him. Perhaps a key phrase, to point up the difference between us, is his description of the transformed ape *Homo divinus,* as "biologically unchanged but spiritually distinct" (p. 31). For me *Homo* is biologically *changed* and spiritually *developed* (sadly only very slightly developed, but that is a different point on which I suspect we would not disagree!). The fact that "Genesis 1 describes humankind as being made in God's image, by a specific act of God" indicates to me nothing more significant than that the thinking of religious visionaries 2,600 years ago was very different from that of a twenty-first century biologist. I cannot embody it in my anthropology.

Victorian Christian reactions

This is a perfect point at which to turn to John Hedley Brooke's thoughtful account of the wildly varying Christian reactions to Darwin in the years following 1859. John points up the long history of contention over the "deeply divisive issue" – about which Darwin himself continually vacillated – "whether nature could be assigned an autonomy and integrity such that divine intervention was unnecessary (even demeaning to the deity), or whether Christian theology had such an investment in the miraculous that to outlaw the possibility of divine incursion in the world would be incongruous" (p. 56). As will already be apparent, I myself am firmly on the side of natural autonomy. And one of the judgements from the generation following Darwin which has been most widely quoted in the past 25 years is that of Aubrey Moore (1891): "The one absolutely impossible conception of God, in the present day, is that which represents him as an occasional visitor. … Either God is everywhere present in nature, or He is nowhere."[4] Again (though not quoted directly in John Brooke's paper here):

> The scientific evidence in favour of evolution, *as a theory*, is infinitely more Christian than the theory of 'special creation'. For it implies the immanence of God in nature, and the omnipresence of His creative power. Those who oppose the doctrine of evolution in defence of a 'continued intervention' of God, seem to have failed to have noticed that *a theory of*

[4] Aubrey L. Moore, "The Christian Doctrine of God," in *Lux Mundi*, ed. C. Gore (London: Murray, 1891).

> *occasional intervention implies as its correlative a theory of ordinary absence.* [5]

The pertinence of this to my comments on Sam Berry's position will be obvious. Yet it is a cautionary observation, typical of John, that Darwin himself was greatly exercised by the cruelty of the natural world – of "nature red in tooth and claw," as Tennyson had so vividly put it a few years before *The Origin* was published. A God who is everywhere present in nature appears to have little concern for creaturely suffering. No-one saw this more clearly than Charles Darwin.

John Brooke gives less space here than he has in other writings to the impact of *Essays and Reviews* (published 1860),[6] but whenever I think about the *The Origin* I find it salutary to remember that, when both volumes were simultaneously on the booksellers' shelves, *Essays* sold three times more copies. For most British readers, *Essays* provided the first exposition of the application of literary and historical analysis to the Bible – a discipline which had been developing in Germany for the previous thirty years, and had become known as the Higher Criticism. The realisation, for instance, that Genesis was a relatively late contribution to Hebrew religious literature shocked many Victorians more than Darwin's account of the mechanism of evolution. And so it should have done!

Expanding on that point, the past few years of reading, writing and speaking around the Darwin anniversaries have brought home to me, more vividly than ever before, the value of really immersing oneself in the intellectual history of such a formative period. And no two writers have done more to enlarge my understanding of it than John Brooke and David Knight. In the perspective of two centuries the present-day attitudes of both the Creationists and the New Atheists, of the tabloid press and, too often, the man or woman in the pew, come to seem not merely infuriating but myopically blinkered and jejune. To see this clearly, it is necessary not just to note that "X wrote so-and-so in NNNN," but to soak oneself as far as possible in the situation of the time – its knowledge and its ignorance, its confidence and its fears, its clarities and its confusions. One must strive not to look superficially at the past like a tourist off a cruise ship, but immerse oneself in it as a good anthropologist does with a society s/he is studying. We must delve, as David says, "until we hear the voices" (p. 78).

---

[5] Aubrey L. Moore, *Science and Faith* (London: Kegan Paul, Trench & Co., 1889).

[6] Benjamin Jowett et al., *Essays and Reviews* (London: Longman, Green, Longman & Roberts, 1860).

Natural Theology

David Fergusson's response to past voices leads him to a taxonomy of different types of natural theology, which he subsequently applies to the debates set in train by *The Origin*, and Darwin's later writings – including a number of topics I have mentioned above. Let me stress his emphasis that the key distinction between different viewpoints is whether or not they share "the assumption that science and religion occupy common ground in their explanatory accounts of nature" (p. 86). Paley assumed this, and Richard Dawkins is therefore right to claim that the Paleyan watchmaker was blind. The Creationists assume it too – and so do far, far too many of our neighbours in the pews and even still too many pulpits. To my mind it is the crux of all Science/Religion debate since the Enlightenment, if not since the Renaissance. Naïve religion does make that assumption; mature religion cannot do so.

To my mind, as I think to David Fergusson's, there are two ways in which the religious mind can relate to scientific world-views while avoiding that assumption. The first is an essentially deistic way: God created the world, but does not intervene in it: the laws once set up suffice. We can answer Stephen Hawking that way: Hawking's equations may describe the initial singularity, but God it was who had caused the equations themselves to operate. Similarly, once life exists, and heritable variation occurs, natural selection will result in the differentiation of varieties. Instance the words David quotes from Frederick Temple: "God did not make the things, we may say: no, but He made them make themselves" (p. 88). General providence takes over from special. However, the deficiency in this approach is the one evident to Aubrey Moore, that the God thus described is too distant to be meaningful. Moore wanted an ongoing view of divine action, and encapsulated it in the concept of immanence. David Fergusson prefers to refer to complementarity between scientific and theological accounts – a concept to which Sam Berry also alludes favourably late in his paper (p. 38). I shall return to this theme at the end of this overview. For the present it is sufficient to note how very different both immanence and complementarity are from deism.

I now have the pleasure of looking (too briefly of course, as with all the others) at Sarah Coakley's punchy comment. First, let me assure her that, though writing from Scotland, I have no desire at all to genuflect towards Karl Barth. As far as I am concerned, if he had never written, Christianity would be better off. Furthermore I agree emphatically that the flat-plane model must not be accepted. The statement "God created … every winged fowl after his kind" (Genesis 1:21) should not be regarded as being of the same sort as "birds evolved from reptiles via intermediate forms such as gliding lizards." The two statements are compatible precisely because they are *not* on the same plane. I am pleased to think that Sarah agrees with this. It is a point which will be taken forward at the end of this essay.

Intelligent Design

Denis Alexander's essay comes next. This is not the paper he gave at the September 2009 meeting, because that was already promised elsewhere. To my mind, however, it is an even more penetrating and damning critique of ID, if that be possible. I am bemused to ask yet again what it is that drives intelligent people such as Behe, Demski and Meyer to invoke an agent (they carefully avoid the word 'God') who from time to time stirs the pond she/he/it has created, in order to overcome the inadequacies of her/his/its initial efforts? How infinitely more worship-worthy is a being who got it right from the beginning!

To me, however, a key remark of Denis's is that "Christians, of all people, have no hidden investments in scientific ignorance." (p. 126). My own disposition is to extend that comment to those of any religious inclination, not Christians only, for the point is of the absolute essence of the confrontation between those of a faith and those of none. But Denis is doubtless implying recognition that Christianity's uniquely incarnational stance gives it a particular investment in judicious understanding of the relationship between God and matter. Perhaps partly for that very reason – but also for the surely-related one that science has developed most in Christendom – the Christian faithful have all too often, over many centuries, foolishly seized upon scientific ignorance to justify their own positions. Always they have been wrong to do so. It was the Oxford theoretical chemist and Methodist, Charles Coulson, who coined the wonderful phrase for this: the 'God-of-the-gaps' argument.[7] Such an argument is tactically injudicious because a scientific gap will always be plugged in the fullness of time. But, more profoundly, it is strategically wrong because it misplaces divine action. As Denis goes on to emphasize, and as we have already noted that Aubrey Moore saw more than a century ago, the God-world relationship must be seen as an immanent sustaining, an ongoing process in the divine mind, not an occasional intervention to make good an original design-insufficiency. We recognized above that this

[7] Charles A. Coulson, *Science and Christian Belief* (London: Fontana, 1960).

viewpoint brings the problem of evil into particularly harsh perspective; but in any case evil is a matter which ID seems little disposed to address.

One more thought: ID, as the term is currently applied, limits itself essentially to problems of biochemistry – to structures and processes which it claims, rightly or wrongly, are not at present understood. Yet I wonder whether the same temptation, to attribute to God processes which we do not currently understand, is not yielded to at least as widely in the context of psychology? We are less explicit, because our knowledge of the mind is so much less precise than our knowledge of molecules. But do not many people think, for instance, of freewill as a phenomenon of something akin to what the ID theorists call 'irreducible complexity'?

At this point in the original conference Sjoerd Bonting responded to Denis Alexander's then paper. As it was a different paper than the one now before us, Sjoerd's response is not printed here. However, I did note at the time a comment which remains pertinent. Observing that religious thinkers who reject ID often regard as a valuable alternative the concept of 'Theistic evolution', in which every mutation is controlled by God, Sjoerd asked: "If that is the case, why are so many of the mutations deleterious?" He preferred to think of mutation as a truly random process, exactly as an atheist would do, with the implied divine self-restraint being in his eyes an instance of kenosis.

Semiotics

I now turn to this book's editor, Andrew Robinson, and his co-author Christopher Southgate. As one who has never got on with Peirce's thought, I have first to admire their paper simply for its success in making constructive use of the opaque American's ideas. But an approach which can produce both a scientific hypothesis about pre-biotic evolution and a theological exegesis of Trinitarian doctrine must be respected for more than simply its grasp of a difficult earlier writer's work. I do not myself feel capable of taking it further, but am sure that the two authors will do so in further writings. Furthermore, their placing of their own thought, in relation to both David Fergusson's typology and Sarah Coakley's comments, admirably fulfils one of the functions that might otherwise be expected of an overviewer!

A reader who clearly does empathize with Peirce's thought is Kenneth Wilson, and probably my best service to the Robinson / Southgate paper will be to highlight a couple of Kenneth's summaries. Firstly: "The linking of the emergence of meaning-making and meaning-seeking to the evolutionary process is of profound significance. … [The paper shows] just how the development of cognition and the ability to 'recognize' and interpret signs takes place progressively over the whole period of human evolution. Our growing self-awareness is the product of a growing capacity to interpret signs" (p. 150). And then: "God, as Trinity, is wholly other (the Father), wholly embodied (the Son), and wholly present (the Holy Spirit). … No separation here, but no simple identity either: rather a coherent eternal relationship." (pp. 150 & 152).

Contributed papers

Last, among the papers preceding this one, are the three contributed by conference delegates. I have already revealed my distaste for the writings of Karl Barth, but must concede that Philip Chapman makes a sustainable case that "Barth and the Darwinists can talk." He does not convince me that this is possible, let alone worthwhile, but *chacun a son gout.*

Teilhard is, at first thought, more appealing to an evolutionist: "Evolution is the biological mechanism by which God is in process of creating the world and moving deeper within it" (p. 166). But Michael Burdett is frank about the most basic of the problems: Bergsonian obscurantism is useless to a twenty-first century biologist, whereas Darwin's clean and simple insight is the basis of all that such a biologist does. Nevertheless, Teilhard's 'within' vividly recalls the 'inscape' and 'thisness' of Gerard Manley Hopkins, briefly but tellingly referred to by Andrew and Christopher (p. 143). Teilhard's Christocentric eschatology must surely also appeal to them? Yet I am left with one non-trivial difficulty: given that "it is precisely the social and technological nature of humanity that is the prime locus in determining humanity's future" (p. 168), is this really compatible with the metaphysical/mystical proclamation that the 'noosphere' is evolving ineluctably towards its 'Omega Point'? Are the two claims simply expressed in different linguistic styles, or do they indicate a deeper mismatch? Technology as an agent of metaphysics? Perhaps, but I need convincing.

It is a privilege to come finally to Fabien Revol's seven gifts of Darwin to theology. Fabien manages, in just a few pages, to outline a complete theology of evolution. He draws constructively on the thought of Arthur Peacocke, which must commend him to any member of the Science and Religion Forum which, but for Arthur, would almost certainly not exist. In particular, he notes early on that "biological evolution is not exhausted by the 'mutation/selection' model, but rather includes also an ethologic approach … which emphasizes the intrinsically dynamic character of the created order" (p. 173). I shall argue shortly that the 'mutation/selection' model has been extended in the last generation by more than ethology; but ethology is an acceptable place to start. Equally welcome is his contention, following Ian Barbour, that "design is not to be found in the details of nature but in general tendencies."

The comparison between laws of nature and laws of behaviour has tempted many writers more established than Fabien yet is, but I remain unconvinced that it is helpful. 'Laws of nature' are human generalizations about the way the world functions, 'laws of behaviour' are moral injunctions; to my mind the use of the same word for both is a category mistake. But I am happy to see him invoke John Haught's view of God as compassionately involved in his creatures' suffering – the Patripassian heresy in mediaeval eyes, but compelling to many of us moderns. Among these was A.N. Whitehead, but his concept of apparently inanimate matter as quasi-biological, rightly seen by Fabien as linking him to Bergson, for that very reason makes him suspect to a twenty-first century scientist, even though the dynamic aspects of both their outlooks (p. 173) is itself more than acceptable. In the end, however, Fabien comes back to "the Tractarians of the Oxford Movement," in particular to Aubrey Moore (p. 178). Moore seems to be the past author most quoted in this book, apart from Charles Darwin. Nuanced now by kenosis, and God's involvement in suffering, his immanentism provides an admirably representative concept with which to conclude discussion of the conference contributions.

I have four points of my own to add.

Modern evolutionary thought

Firstly, it would be inappropriate to conclude a 2012 volume on evolution and theology without acknowledging how much more than random mutation is now recognized as influencing the form – the 'phenotype' – of any plant or animal. Inescapable physical forces (including some which tend toward unpredictable order [8]), massive constraints upon embryo formation, and ecological factors operating prior to natural selection (hinted at in the last paper),[9] are one category. Another is the vast complexity of the genome itself, in no way suggested by Mendel's observations, and their molecular maturation in the work of Watson and

---

[8] Stuart A. Kauffman, *The Origins of Order: Self-Organization and Selection in Evolution* (New York & Oxford: Oxford University Press, 1993).

[9] Sean B. Carroll, *Endless Forms Most Beautiful: The New Science of Evo-Devo* (London: Phoenix, 2006).

Crick. The majority of DNA is actually 'junk', in that 'bean-box' genetic perspective: that is to say, it does not code for proteins, and so get expressed directly in features of the phenotype. Instead, it controls the operation of those direct-action genes.[10] The current overall picture is colloquially designated 'eco-evo-devo' – the interaction of this broadened genetics with developmental biology, intimate environment, wider ecology and individual history, to result in the free-living form which is subject to selection.

Darwin was not wrong. Only fools and self-publicists claim that he was. There are many hints in *The Origin* and other writings that he was already aware in principle of many of these kinds of influence; he simply had almost no specific information to write about. In any case, natural selection remains the final arbiter. The last paragraph of one of the great modern textbooks encapsulates the current professional perspective:

> Evolution by natural selection, the mainstay of the Darwinian synthesis, is not in question. But we have realised that animals have been able to evolve in many different ways. Most of this diversification involves the mechanisms used for the generation of variation, and that is where development plays its enormous roles. Natural selection works on the array of variants that development offers it. Evolution by natural selection isn't going away. It's just getting a whole lot richer.[11]

If there is a theological response to this re-weighting of the biological account it must surely be a like re-weighting – an adjustment of degree, not kind. Natural selection remains, but the variations upon which it is working are less dominated by the apparent randomness of gene mutation; rather, the mature being (particularly the human being) is to be seen as ever more deeply entwined and intervolved with the physical and biological worlds from which it, she or he has sprung. Divinity is ever more radically indwelling, but indwelling intimately through the material world which is that divinity's creation, not as a super-addition to it.

Divine involvement

My second point is to take rather further than any contributor has so far the question of how to relate God's action to the evolutionary events

---

[10] Gabriel A. Dover, *Dear Mr Darwin: Letters on the Evolution of Life and Human Nature* (London: Phoenix, 2001).

[11] Scott F. Gilbert and David Epel, *Ecological Developmental Biology* (Sunderland, MA: Sinauer, 2009).

observed by biologists. Sarah Coakley comes nearest. As she says (pp. 101-102), in challenging the complacency of atheistical neo-Darwinists,

> … and to destabilize the flat plane that they still believe themselves to enjoy commanding … [t]he first move, the ontological move, is … the classic Thomist one … that whatever God's actions involve they are operative at a completely different ontological level than that of discrete created events. This position already complicates the flat-plane position by insisting that it is not the only, or finally undergirding, scene of operation. … God does not compete for space with individual events studied by the scientist.

To take this further, John Polkinghorne looks to chaotic events, and many others look to quantum ones, as loci in which God could act without our being able to detect the intervention. If obliged to choose between these, I would opt for the quantum level; there, our ignorance has, since Bohr and Heisenberg worked together in Copenhagen, been considered to be absolute and inescapable – an ontological indeterminacy. By contrast, no single step in a chaotic sequence is beyond scientific understanding, we are merely overwhelmed by the near-infinite mass of information which would be required to predict an outcome: the indeterminacy is epistemological but not ontological, and so to my mind less ultimate (though I know Sir John disagrees). But both chaos and quantum arguments depend on current scientific understanding, an unstable position which seems not to concern ID proponents but greatly concerns me. As William Temple said somewhere: "A religion which marries the science of one generation will be a widow in the next."

Instead I turn to the fact that a given system or structure can often be described at a number of different levels. There are endless examples within my own field, physiology, but probably more generally familiar, to an extent sufficient for the present purpose, are the operations of a personal computer. The most remote from one another – the most nearly independent or 'dissociated' – are the account which can be given in terms of solid-state electronics, with its movements of electrons and positive holes between semiconductor devices, and the account which would be given by the computer's user, typing her article or setting out a spreadsheet. The changed states of the transistors and other semiconductor units caused by the user's key-strokes would have to be treated as random in the electronic account, for they would have no explanation at that level. Thus the two accounts would make no mention one of another, but would not be strictly independent: they would be nearly, but not absolutely, dissociated. In the theological situation, of course, it is God who is envisaged as striking the keys, and the events analogized by the electronic account might, for our present purposes, be mutations in DNA; these would be analysable to the Nth degree by a molecular geneticist, without offering the least evidence that God had triggered the events. Those who believe we can identify the locus of God's actions in the world would say that the computer analogy is exact: that the divine level and the scientifically-observable level were almost, but not totally, dissociated – that a coupling we can comprehend in principle actually causes events which seem random when viewed from the electronic-circuit or DNA-mutation level. However, in my view it is just as feasible that, in the case of divine action, the dissociation may be absolute: that no mechanism whatsoever, coupling the two, will ever be even theoretically definable by human minds. I like to think that this, expressed in very different language, is what Sarah Coakley means in the passage quoted above.

Karen Armstrong has argued persuasively, particularly in *The Case for God,* [12] that it is an aberration of post-Enlightenment thinking which leads us even to consider the possible mechanism of God's actions in the world. She points out that in Vedic, Confucian and early Buddhist thought, in all mystical traditions, and in pre-Renaissance expressions of the three Abrahamic faiths, the idea that we might aspire to such understanding would never have been formulated. Following her lead, I have lately delved a little into the writings of the Pseudo-Dionysius. In that author's *Mystical Theology* the religious thinker is urged to recognize that theology has ultimately:

> neither word nor act of understanding, since it is on a plane above all this, and is made manifest only to those who travel … beyond the summit of every holy ascent, who leave behind them every divine light, every voice, every word from heaven, and who plunge into the darkness where … there dwells the One who is beyond all things. [13]

This is apophatic theology, the theology which recognizes the futility of words and the concepts those words express. We are still, in the twenty-first century, happy to sing, "Immortal, invisible, God only wise / In light inaccessible hid from our eyes." We are still prepared to describe divinity as 'ineffable'. But can we retrieve the humility really to accept what that word means, and to embody it at the core of our thought?

---

[12] Karen Armstrong, *The Case for God: What Religion Really Means* (London: The Bodley Head, 2009).

[13] Paul Rorem, *Pseudo-Dionysius: a Commentary on the Texts and an Introduction to Their Influence* (New York & Oxford: Oxford University Press, 1993).

What can evolved minds know of God?

My third contention is that, if we accept that human beings are entirely the product of evolution on the surface of Planet Earth, their minds as much as their bodies must be so regarded. In that case we have to ask what relation our God-concepts can have to any underlying, non-temporal and non-material reality. My own answer is: "None of which we can be certain." Our faculties, and the concepts they can formulate, are the products of evolution and the outcomes of natural selection. Consequently those concepts with direct survival-value are highly trustworthy, but those with no such register are so far beyond verification that even their validity as concepts is in doubt. There is scope for a rigorous philosophical consideration of the very meanings we can possibly attach to terms purporting to refer beyond the spatio-temporal world, but even if I were competent to attempt such an analysis it would be far beyond the scope of this chapter. However, the Austrian theoretical biologist Rupert Riedl is perhaps less pessimistic:

> Evolutionary principles of self-organization allowed a post-stabilized harmony to develop, producing sense and purpose within creatures and allowing even God to be revealed or sensed as a necessary hope.[14]

Even if Reidl is right, I find in these considerations further powerful arguments against metaphysical theological constructs of any sort, and hence against dogmatic credal systems which embody them: arguments, in fact, for a maximally apophatic theology once again. However, I have developed this case more than once previously, not least in a previous volume in this series,[15] so I will not dwell further on it here.

---

[14] Rupert Reidl, "A Systems Theory of Evolution," in *Darwinism and Philosophy*, ed. Vittorio Hösle and Christian Illies (Notre Dame, IND: University of Notre Dame Press, 2005). Quoted by Conor Cunningham, *Darwin's Pious Idea: Why the Ultra-Darwinists and Creationists Both Get it Wrong* (Grand Rapids, MI & Cambridge, UK: W.B. Eerdmans, 2010).

[15] Neil. Spurway, "What can Evolved Minds Know of God? Reconsidering Theology in the Light of Evolutionary Epistemology" in *Theology, Evolution and the Mind*, ed. Neil Spurway (Newcastle: Cambridge Scholars Publishing, 2009); "Theology's Response to Evolutionary Epistemology. What can Evolved Beings Know of God?" in *Linnaeus and Homo Religiosus. Biological Roots of Religious Awareness and Human Identity*, ed. Carl R. Bråkenheilm (Uppsala: Uppsala Universitet, 2010).

Referring to the discussion between David Fergusson and Sarah Coakley (Chapters 6 and 7) these considerations suggest to me that no theology which is not natural can be valid! However, I leave this as a thought for the reader to conjure with as (s)he wishes.

Finally

Surveying this book, we have found Sam Berry happy to sustain a conservative evangelical stance in parallel with evolutionary science. How would the other contributors interpret the mechanism of what Andrew Robinson and Christopher Southgate refer to as "God's gracious self-revelation," or the chemistry of incarnation, or the physics of resurrection? One suspects that most, or all, of the invited contributors would offer sophisticated, non-literal accounts of these phenomena or adopt something close to my own apophatic stance. On the conference floor, however, there were people who would clearly wish to think of revelation as some form of divine injection into the scientifically-describable world. And similarly for incarnation, if not resurrection. Darwin himself would, I suspect, have been intrigued by this diversity, but more than comfortable with what this wide range of opinion-holders have in common. As late as 1879 he wrote to a correspondent:

> It seems to me absurd to doubt that a man may be both an ardent Theist and an Evolutionist. ... In my most extreme fluctuations I have never denied the existence of God. [16]

Nevertheless, if the approach I have outlined is on anywhere near the right lines, Darwinian thinking keeps us earth-bound. We are animals, albeit particularly brainy animals, and are wholly the products of terrestrial evolution. What this does *not* say, however, is what it is all too widely assumed to say, namely that humanity is diminished by this assessment. Precisely as logical as diminution would be to say that the other species, and the world in which they and we have come to be, are elevated, enriched, enchanted, made holy by being seen as at one with *imago dei.*

We are "of the earth, earthy."[17] What glorious things that says about the earth!

---

[16] Letter to John Fordyce, 7th May 1879.

[17] 1 Corinthians 15: 47

BIBLIOGRAPHY

Alexander, Denis R. *Creation or Evolution - Do We Have to Choose?* Oxford: Monarch, 2008.

—. "Is Intelligent Design Intelligent?" In *Theology After Darwin*, edited by R. J. Berry and M. Northcott. Milton Keynes: Paternoster, 2009.

—. *The Language of Genetics - an Introduction*. Philadelphia: Templeton Foundation Press, 2011.

Arndt, William F. and F. Wilbut Gringrich, *A Greek-English Lexicon of the New Testament and other Early Christian Literature*, 4th revised and augmented edition. Chicago: Chicago University Press, 1952.

Armstrong, Karen. *The Case for God: What Religion Really Mean*s. London: The Bodley Head, 2009.

Axe, Douglas D. "Estimating the prevalence of protein sequences adopting functional enzyme folds." *Journal of Molecular Biol*ogy 341 (5) (2004):1295-315.

Axe, Douglas D., B.W. Dixon, and P. Lu. "Stylus: a system for evolutionary experimentation based on a protein/proteome model with non-arbitrary functional constraints." *PLoS One* 3 (6) (2008): e2246.

Ayala, Francisco. *Darwin's Gift to Science and Religion.* Washington: Joseph Henry Press, 2007.

Bailey, Edward. *Charles Lyell*. Edinburgh: Nelson, 1962.

Balfour, Arthur J. *A Defence of Philosophic Doubt; being an Essay on the Foundations of Belief.* London: Macmillan, 1879.

—. *Foundations of Belief: being Notes Introductory to the Study of Theology,* 2nd edition. London: Longmans, Green, 1895.

Barbour, Ian G. *Religion and Science, Historical and Contemporary Issues*. New York: Harper San Francisco, 1997.

Barlow, Nora, ed. *The Autobiography of Charles Darwin 1809-1882*. London: Collins, 1958.

Barr, James, *Biblical Faith and Natural Theology.* Oxford: Clarendon Press, 1993.

Barrett, Justin L. *Why Would Anyone Believe in God?* Lanham MD: AltaMira Press, 2004.

Barrick, J.E., D.S. Yu, S.H. Yoon, H. Jeong, T.K. Oh, D. Schneider, R.E. Lenski, and J.F. Kim. "Genome evolution and adaptation in a long-term experiment with Escherichia coli." *Nature* 461 (7268) (2009): 1243-7.

Barth, Karl. *Church Dogmatics.* Translated and edited by G.W. Bromiley, T.F. Torrance and others. Edinburgh: T&T Clark, 1937-1975.

Bartholomew, David J. *God of Chance*. London: SCM, 1984.

Bartholomew, David. *God, Chance and Purpose*. Cambridge: Cambridge University Press, 2008.

Beer, Gillian. *Darwin's Plots: Evolutionary Narrative in Darwin, George Eliot and Nineteenth-Century Fiction.* London: Routledge and Kegan Paul, 1983.

Bellon, Richard, "Joseph Hooker takes a 'fixed post': transmutation and the 'present unsatisfactory state of systematic botany', 1844-1860," *Journal of the History of Biology* 39 (2006): 1-39.

Bergson, Henri. *Creative Evolution*. Translated by Arthur Mitchell. New York: Henry Holt, 1911.

Berry, R.J. *Neo-Darwinism.* London: Edward Arnold, 1982.

—. "Ecology: where genes and geography meet." *Journal of Animal Ecology* 58 (1989): 733-59.

—. "This cursed earth. Is 'the Fall' credible?" *Science & Christian Belief* 11 (1999): 29-49, 165-67.

—. ed. *The Care of Creation*. Leicester: IVP, 2000.

—. *God's Book of Works.* London: T&T Clark, 2003.

—. "Adam or Adamah?" *Science & Christian Belief* 23 (2011): 23-48.

Berry, R.J., T.J.S. Crawford, and G.M. Hewitt, eds. *Genes in Ecology.* Oxford: Blackwell Scientific, 1992.

Bimson, John J. "Doctrines of the Fall and Sin After Darwin." In *Theology After Darwin,* edited by Michael S. Northcott and R.J. Berry. Milton Keynes: Paternoster, 2009.

Blocher, Henri. *In the Beginning: The Opening Chapters of Genesis.* Leicester: IVP, 1984.

Bolton, D.C., M.P. McKinley, and S.B. Prusiner. "Identification of a protein that purifies with the scrapie prion." *Science* 218 (4579) (1982):1309-11.

Bowler, Peter J. *The Eclipse of Darwinism.* Baltimore: Johns Hopkins University Press, 1983.

—. *Theories of Human Evolution: a Century of Debates 1844-1944*. Oxford: Basil Blackwell, 1986.

—. "Evolution and the Eucharist: Bishop E.W. Barnes on science and religion in the 1920s and 1930s," *British Journal for the History of Science* 31 (1988): 453-67.

—. *Reconciling Science and Religion: The Debate in Early Twentieth Century Britain.* Chicago: University of Chicago Press, 2001.

Britten, Roy J. "Divergence between samples of chimpanzee and human DNA sequences is 5%, counting indels," *Proceedings of the National Academy of Sciences of the USA* 99 (2002): 13633-35.

Brooke, John Hedley. "The Relations between Darwin's Science and his Religion." In *Darwinism and Divinity*, edited by John R. Durant. Oxford: Blackwell, 1985.

—. "Indications of a Creator: Whewell as Apologist and Priest." In *William Whewell: A Composite Portrait*, edited by Menachem Fisch and Simon Schaffer. Oxford: Oxford University Press, 1991.

—. *Science and Religion: Some Historical Perspectives.* Cambridge: Cambridge University Press, 1991.

—. "Natural Law in the Natural Sciences: the Origins of Modern Atheism?" *Science and Christian Belief* 4 (1992): 83–103.

—. "The Natural Theology of the Geologists: Some Theological Strata." In *Images of the Earth: Essays in the History of the Environmental Sciences,* 2nd edition, edited by L. J. Jordanova and Roy Porter. Oxford: Alden Press for the British Society for the History of Science, 1997.

—. "The Wilberforce-Huxley debate: why did it happen?" *Science & Christian Belief* 13 (2001): 127-141.

—. "Genesis and the Scientists: Dissonance among the Harmonizers." In *Reading Genesis after Darwin,* edited by Stephen C. Barton and David Wilkinson. Oxford: Oxford University Press, 2009.

—. " 'Laws impressed on matter by the Creator'?: The *Origin* and the Question of Religion." In *The Cambridge Companion to the "Origin of Species,"* edited by Michael Ruse and Robert J. Richards. Cambridge: Cambridge University Press, 2009.

Brooke, John, and Geoffrey Cantor. *Reconstructing Nature: The Engagement of Science & Religion.* Edinburgh: T & Clark, 1998; New York: Oxford University Press, 2000.

Brown, Warren S., Nancey Murphy, and H. Newton Malony, eds. *Whatever Happened to the Soul?* Minneapolis: Fortress Press, 1998.

Browne, Janet. *The Secular Ark: Studies in the History of Biogeography.* New Haven: Yale University Press, 1983.

—. *Charles Darwin: Voyaging.* London: Jonathan Cape, 1995.

Burkhardt, Frederick, et al., eds. *The Correspondence of Charles Darwin*, 18 vols. Cambridge: Cambridge University Press, 1985-2010.

Burkhardt, Frederick, and Sydney Smith. *The Correspondence of Charles Darwin, Volume 4 1844-1846.* Cambridge: Cambridge University

Press, 1988.

Carroll, Sean B. *Endless Forms Most Beautiful: The New Science of Evo-Devo*. London: Phoenix, 2006.

Chalmers, Thomas. *On the Power, Wisdom, and Goodness of God as Manifested in the Adaptation of External Nature to the Moral and Intellectual Constitution of Man, 2 vols*. London: 1833.

Chambers, Robert. *Vestiges of the Natural History of Creation*. London: John Churchill, 1844.

Chung, Paul, S. "Karl Barth and God in Creation: Towards an Interfaith Dialogue with Science and Religion." *Theology and Science* 3 (2005): 56-69.

Churchland, Paul M. "Eliminative Materialism and the Propositional Attitudes," in *Mind and Cognition,* edited by William G. Lycan, 120-132. Malden, Mass. and Oxford: Oxford University Press, 1999.

Clayton, Philip. *Mind and Emergence: from Quantum to Consciousness*. New York: Oxford University Press, 2005.

Cobb Jr, John B. *A Christian Natural Theology*, 2nd edition. Louisville: Westminster John Knox Press, 2007.

Collingwood, R.G. *Speculum Menti*. Oxford: Oxford University Press, 1924.

Collins, Francis. *The Language of God: A Scientist Presents Evidence for Belief*. New York: Free Press, 2006.

Colyer, Peter. "Kenosis by Addition, Kenosis by Subtraction." In *Humanity, The World and God*, Studies in Science and Theology, vol. 11, edited by Willem B. Drees, Hubert Meisinger and Taede A. Smedes. Lund: University of Lund, 2007-8.

Conway Morris, Simon. *Life's Solution: Inevitable Humans in a Lonely Universe*. Cambridge: Cambridge University Press, 2003.

—. "The Boyle Lecture 2005: Darwin's compass." *Science & Christian Belief* 18 (2006): 5-22.

—. ed. *The Deep Structure of Biology. Is Convergence Sufficiently Ubiquitous to Give a Directional Signal?* West Conshohocken, PA: Templeton Foundation Press, 2008.

Cooper, Nicholas, ed. *John Ray and his Successors: the Clergyman as Biologist*. Braintree: John Ray Trust, 2000.

Coulson, Charles A. *Science and the Idea of God*. Cambridge: Cambridge University Press, 1958.

—. *Science and Christian Belief*. London: Fontana, 1960.

Cranfield, Charles. "Some observations on Romans 8: 19-21." In *Reconciliation and Hope: New Testament Essays on Atonement and Eschatology presented to L.L. Morris on his 60th Birthday*, edited by

R. Banks. Grand Rapids: Eerdmans, 1974.
Cunningham, Conor. *Darwin's Pious Idea: Why the Ultra-Darwinists and Creationists Both Get it Wrong*. Grand Rapids, MI & Cambridge, UK: W.B. Eerdmans, 2010.
Darwin, Charles. *On the Origin of Species by Means of Natural Selection, or the Preservation of Favoured Races in the Struggle for Life*. London: John Murray, 1859.
—. *The Variation of Animals and Plants under Domestication*. London: Murray, 1868.
—. *The Decent of Man and Selection in Relation to Sex*. London: John Murray, 1871.
—. "Darwin's Notebooks on Transmutation of Species, Part I. First Notebook (July 1837 - February 1838). Edited with and Introduction and Notes by Gavin De Beer," *Bulletin of the British Museum (Natural History)* 2 (1960).
—. *The Descent of Man and Selection in Relation to Sex [1871]*. London: Penguin, 2004.
—. *The Origin of Species*. Edited by J.W. Burrows. Harmondsworth: Penguin, 1985. [Contains the text of the first edition]
Darwin, Francis, ed. *The Life and Letters of Charles Darwin*, 3 vols. London: Murray, 1887.
Dawkins, Richard. *The Blind Watchmaker*. London: Longman, 1986.
Day, Matthew. "Godless Savages and Superstitious Dogs: Charles Darwin, Imperial Ethnography and the Problem of Human Uniqueness." *Journal of the History of Ideas* 69 (2008): 49-70.
de Waal, Frans M.B. *Good Natured: The Origins of Right and Wrong in Humans and Other Animals*. Cambridge, MA: Harvard University Press, 1996.
Deacon, Terrence W., "Reciprocal Linkage between Self-Organizing Processes is Sufficient for Self-Reproduction and Evolvability," Biological Theory 1 (2006): 136-149.
Deacon, Terrence W. *The Symbolic Species*. Harmondsworth: Penguin, 2007.
Deane-Drummond, Celia. *Christ and Evolution: Wonder and Wisdom*. Minneapolis: Fortress; London: SCM Press, 2009.
Deane-Drummond, Celia and David Clough. *Creaturely Theology: On God, Humans and Other Animals*. London: SCM Press, 2009.
Dembski, William A., and Michael Ruse, eds. *Debating Design: From Darwin to DNA*. Cambridge: Cambridge University Press, 2004.
Dennett, Daniel C. *Darwin's Dangerous Idea*. New York: Touchstone, 1995.

—. *Breaking the Spell*. London: Penguin, 2007.

Desmond, Adrian. *The Politics of Evolution*. Chicago: Chicago University Press, 1989.

Diamond, Jared. *The Rise and Fall of the Third Chimpanzee*. London: Radius, 1991.

Dobzhansky, Theodosius. *Genetics and the Origin of Species*. New York: Columbia University Press, 1937.

—. *Mankind Evolving*. New Haven: Yale University Press, 1962.

Dover, Gabriel A. *Dear Mr Darwin: Letters on the Evolution of Life and Human Nature*. London: Phoenix, 2001.

Drummond, Henry. *The Lowell Lectures on the Ascent of Man*. London: Hodder and Stoughton, 1894.

—. *The Ascent of Man*. London: Hodder and Stoughton, 1904.

Dugatkin, Lee Alan. *The Altruism Equation*. Princeton: Princeton University Press, 2006.

Durant, John, ed. *Darwinism and Divinity*. Oxford: Basil Blackwell, 1985.

Eliot, T.S. *Collected Poems and Plays*. London: Faber and Faber, 1969.

Ellul, Jacques. *What I Believe*. Translated by Geoffrey William Bromiley. London: Eerdmans, 1989.

England, Richard. "Natural Selection, Teleology, and the Logos: From Darwin to the Oxford Neo-Darwinians." *Osiris* 16 (2001): 270-87.

Ewart, Paul. "The Necessity of Chance: Randomness, Purpose and the Sovereignty of God." *Science and Christian Belief* 21 (2009): 111-131.

Fergusson, David. *The Cosmos and the Creator*. London: SPCK, 1998.

—. "Types of Natural Theology." In *The Evolution of Rationality: Interdisciplinary Essays in Honor of J. Wentzel Van Huyssteen,* edited by F. LeRon Shults, 380-393. Grand Rapids: Eerdmans, 2006.

—. "Providence after Darwin'." In *Theology After Darwin,* edited by Michael Northcott and R. J. Berry, 73-88. Milton Keynes: Paternoster: 2009.

Ferrada, E., and A. Wagner. "Evolutionary innovations and the organization of protein functions in genotype space." *PLoS One* 5 (11) (2010): e14172.

Feyerabend, Paul. *Against Method*. London: New Left Books, 1975.

Fichman, Martin. *An Elusive Victorian: the Evolution of Alfred Russel Wallace.* Chicago: University of Chicago Press, 2004.

Finlayson, Clive. "Biogeography and evolution of the genus *Homo." Trends in Ecology and Evolution,* 20 (2005): 457-63.

Fisher, R.A. "The correlation between relatives on the supposition of Mendelian inheritance." *Transactions of the Royal Society of Edinburgh,* 52 (1918): 399-433.

—. "On the dominance ratio." *Proceedings of the Royal Society of Edinburgh*, 42 (1922): 321-41.

—. *The Genetical Theory of Natural Selection.* Oxford: Clarendon Press, 1930.

Ford, E.B. *Mendelism and Evolution.* London: Methuen, 1931.

Francis, Mark. *Herbert Spencer and the Invention of Modern Life.* Stocksfield: Acumen, 2007.

Frei, Hans. *Types of Modern Theology.* New Haven: Yale University Press, 1992.

Gabellieri, Emmanuel. "Nature et création: Opposition ou nouvelle alliance ?" In *Nature et création entre science et théologie,* edited by J.M. Exbrayat, E & Gabellieri, 181-230. Paris: Librairie Philosophique J.Vrin-IIEE, 2006.

Giberson, Karl W. *Saving Darwin*. New York: HarperCollins, 2008.

Gilbert, Scott F. and David Epel. *Ecological Developmental Biology.* Sunderland, MA: Sinauer, 2009.

Gould, Stephen Jay. *Wonderful Life*. New York: Norton, 1989.

Gray, Asa. *Darwiniana*. Edited by A. Hunter Dupree. Cambridge: Harvard University Press, 1963.

Gregory, Frederick. "The Impact of Darwinian Evolution on Protestant Theology in the Nineteenth Century." In *God and Nature: Historical Essays on the Encounter between Christianity and Science,* edited by David C. Lindberg and Ronald L. Numbers. Berkeley: University of California Press, 1986.

Habermas, Jürgen. *The Theory of Communicative Action: Vol. 2 Lifeworld and System, A Critique of Functionalist Reason.* Translated by Thomas McCarthy. Cambridge: Polity, 1987.

Haldane, J.B.S. *The Causes of Evolution.* London: Longmans, 1932.

—. "Population Genetics." *New Biology* 18(1955): 34-51.

Hall, Amy Laura. "Charles Kingsley's Christian Darwinism." In *Theology After Darwin,* edited by Michael S. Northcott and R.J. Berry. Milton Keynes: Paternoster, 2009.

Hamilton, W.D. "The genetical evolution of social behaviour." *Journal of Theoretical Biology* 7 (1964): 1-52.

Hartshorne, Charles. *Omnipotence and Other Theological Mistakes.* New York: State University of New York, 1984.

Haught, John F. *Deeper than Darwin: The Prospect for Religion in the Age of Evolution.* Boulder, Colorado: Westview Press, 2003.

Haught, John. "The Boyle Lecture 2003: Darwin, design and the promise of nature." *Science & Christian Belief* 17 (2005): 5-20.

—. *God after Darwin, A Theology of Evolution, Second Edition*. Boulder: Westview Press, 2008.

Hawking, Stephen W. *A Brief History of Time – from Big Bangs to Black Holes*. London: Bantam Press, 1988.

Hodge, Charles. *What is Darwinism?* London: Nelson, 1874.

Hopkins, Gerard Manley. *Poems and Prose of Gerard Manley Hopkins*, edited by W.H. Gardner. Harmondsworth: Penguin.

Humphrey, Nicholas. *Seeing Red*. New Haven and London: Belknap Press at Harvard University Press, 2006.

Hunsinger, George. *How to read Karl Barth.* New York and Oxford. Oxford University Press, 1991.

Huxley, Julian. *Evolution: The Modern Synthesis*. London: Allen & Unwin, 1942.

Huxley, Leonard. *The Life and Letters of Thomas Henry Huxley,* Volume 1. London: Macmillan, 1900.

Huxley, Thomas H. "Review of the *Origin of Species*". *Westminster Review*, n.s., 17 (1860): 541-70.

—. *Man's Place in Nature*. London: Williams & Norgate, 1863.

—. "On the Reception of the *Origin of Species*" In *The Life and Letters of Charles Darwin*, *vol. 2*, edited by Francis Darwin. London: Murray, 1887.

—. "An apologetic irenicon." *Fortnightly Review*, n.s., no. 52 (1892): 557-61.

—. "Mr. Gladstone and Genesis." In *Science and Hebrew Tradition: Essays by Thomas H. Huxley*. London: Macmillan, 1904.

Illingworth, John R. "The Incarnation in Relation to Development." In *Lux Mundi: a Series of Studies in the Religion of the Incarnation,* 15th edition, edited by Charles Gore, 132-57. London: Murray, 1899.

Iverach, James. *Evolution and Christianity*. London: Hodder & Stoughton, 1894.

James, Wendy. *The Ceremonial Animal: A New Portrait of Anthropology.* Oxford, Oxford University Press, 2003.

James, William. *The Will to Believe and Other Essays in Popular Philosophy.* New York: Longmans, Green and Co., 1899.

Jeeves, Malcolm. "Towards a composite portrait of human nature." In *From Cells to Souls – and Beyond,* edited by Malcolm Jeeves, 233-49. Grand Rapids: Eerdmans, 2004.

Jones, Steve. *Darwin's Island: The Galapagos in the Garden of England.* London: Little Brown, 2009.

Jowett, Benjamin et al. *Essays and Reviews*. London: Longman, Green, Longman & Roberts, 1860.

Kauffman, Stuart A. *The Origins of Order: Self-Organization and Selection in Evolution*. New York & Oxford: Oxford University Press, 1993.

Kellogg, Vernon L. *Darwinism Today*. London: George Bell & Sons, 1907.

Kerr, Fergus. *After Aquinas: Versions of Thomism*. Oxford: Blackwell, 2002.

Kidner, Derek. *Genesis*. London: Tyndale, 1967.

Kingsley, F.E., ed. *Charles Kingsley: His Letters and Memories of his Life*. London: Kegan Paul, 1883.

Knight, David. *Science and Spirituality: the Volatile Connection*. London: Routledge, 2004.

Larson, Edward J. *Summer for the Gods: The Scopes Trial and America's Continuing Debate over Science and Religion*. Cambridge: Harvard University Press, 1997.

Lewis, C.L.E., and S.J. Knell, eds. *The Age of the Earth: from 4004 BC to AD 2002*. London: Geological Society Special Publication No. 190, 2001.

Lewis, Cherry and Simon Knell, eds. *The Making of the Geological Society of London*. London: Geological Society, 2009.

Leong Ting Lui, Ron Yang, Andrew J.N. Robinson, and Christopher C.B. Southgate, "Interpretation and the Origin of Life," *Biological Theory* 5 (2010): 112-116.

Lightman, Bernard. "Victorian Sciences and Religions: Discordant Harmonies." *Osiris* 16 (2001): 343-66.

—. "Scientists as Materialists in the Periodical Press: Tyndall's Belfast Address." In *Science Serialized: Representations of Science in Nineteenth-century Periodicals,* edited by Geoffrey Cantor and Sally Shuttleworth, 199-237. Cambridge MA; MIT Press, 2004.

—. *Evolutionary Naturalism in Victorian Britain: The 'Darwinians' and their Critics*. Variorum Collected Studies Series. Farnham: Ashgate, 2009.

Lin, Y., L.J. Lin, P. Sriratana, K. Coleman, T. Ha, M. Spies, and I.K. Cann. "Engineering of functional replication protein a homologs based on insights into the evolution of oligonucleotide/oligosaccharide-binding folds." *Journal of Bacteriology* 190 (17) (2008): 5766-80.

Linnean Society of London. *The Darwin-Wallace celebration held on Thursday, 1st July 1908, by the Linnean Society of London*. London: Linnean Society, 1908.

Lipton, Peter. *Inference to the Best Explanation*. London: Routledge, 1991.

Livingston, James C. *Religious Thought in the Victorian Age*. London: T&T Clark International, 2006.

Livingstone, David N., *Darwin's Forgotten Defenders. The Encounter Between Evangelical Theology and Evolutionary Thought.* Grand Rapids: Eerdmans, 1987.

—. *Putting Science in its Place.* Chicago: University of Chicago Press, 2003.

—. "Replacing Darwinism and Christianity." In *When Science and Christianity Meet*, edited by David C. Lindberg and Ronald L. Numbers. Chicago: University of Chicago Press, 2003.

Locke, John. *An Essay Concerning Human Understanding*. 1690.

Lonergan, Bernard. *Insight: A Study of Human Understanding.* New York: Philosophical Library, 3rd edition, 1970.

Lubac, Henri de. *The Religion of Teilhard de Chardin*. Translated by René Hague. London: Collins, 1967.

Macedo, Stephen and Josiah Ober, eds. *Primates and Philosophers: How Morality Evolved*. Princeton, NJ: Princeton University Press, 2006.

MacKay, Donald M. *Science, Chance and Providence.* Oxford: Oxford University Press, 1978.

—. *Behind the Eye*. Oxford Blackwell, 1991.

Mansel, Henry L. *The Limits of Religious Thought,* 4th edition. London: John Murray, 1859.

Mayr, Ernst, and William B. Provine, eds. *The Evolutionary Synthesis*. Cambridge: Harvard University Press, 1980.

McDonald, H.D. *The Christian View of Man*. London: Marshall, Morgan and Scott, 1981.

McGrath, Alister. *The Open Secret: A New Vision for Natural Theology.* Oxford: Blackwell, 2008.

McMullin, Ernan. "Cosmic Purpose and the Contingency of Human Evolution." *Theology Today* 55 (1998): 389-414.

Medawar, Peter B. *The Art of the Soluble*. London: Methuen, 1968.

Messer, Neil. *Selfish Genes and Christian Ethics.* London: SCM, 2007.

Meyer, Stephen C. *Signature in the Cell.* New York: HarperCollins, 2010.

Middleton, J. Richard. *The Liberating Image: The* imago Dei *in Genesis 1*. Grand Rapids: Brazos Press, 2005.

Midgley, Mary. *Evolution as a Religion.* London: Methuen, 1985.

Miller, Gary L. and Robert E. Ricklefs. *Écologie*. Translated from the fourth edition by Michel and Virginie Baguette, Frank d'Amico, and Grégory Mahy. Bruxelles: De Boeck, 2005.

Miller, Hugh. "The Testimony of the Rocks; Or, Geology in its Bearing on the Two Theologies, Natural and Revealed." Reprinted in *Creationism and Scriptural Geology, 1814-1857*, edited by John M Lynch. Bristol: Thoemmes Press, 2002.

Miller, Kenneth R. *Finding Darwin's God*. New York: HarperCollins, 2000.

Monod, Jacques. *Chance and Necessity*. London: Collins, 1971.

Moore, Aubrey L. "The Christian doctrine of God." In *Lux Mundi,* edited by C. Gore, 57-109. London: John Murray, 1889.

—. *Science and the Faith, Essays on Apologetic Subjects, With an Introduction*. 1893. A facsimile of the fourth edition. London: Kegan Paul, Trench, Trübner & Co.

—. "The Christian Doctrine of God." In *Lux Mundi*. 15th edition. Edited by C. Gore, 41-81. London: Murray, 1904.

—. *Science and Faith*. London: Kegan Paul, Trench & Co., 1889.

Moore, James R. *The Post-Darwinian Controversies*. Cambridge: Cambridge University Press, 1979.

—. "Deconstructing Darwinism: The Politics of Evolution in the 1860s." *Journal of the History of Biology* 24 (1991): 353–408.

—. "Ronald Aylmer Fisher: A Faith Fit for Eugenics." In Nicolaas A. Rupke, *Eminent Lives in Twentieth-Century Science and Religion*. Frankfurt: Peter Lang, 2007.

Moore, James, and Adrian Desmond. Introduction to *The Descent of Man*, by Charles Darwin, vii-lviii. London: Penguin Classics, 2004.

Morrell, Jack and Arnold Thackray. *Gentlemen of Science*. Oxford: Oxford University Press, 1981.

Morris, Leon. *The Epistle to the Romans*. Leicester: IVP, 1988.

Moule, C.F.D. *Man and Nature in the New Testament*. London: Athlone, 1964.

Murray, Michael J. *Nature Red in Tooth and Claw: Theism and the Problem of Animal Suffering*. Oxford: Oxford University Press, 2008.

Newberg, Andrew. *Principles of Neurotheology*. Burlington, VT., and Aldershot: Ashgate, 2010.

—. *Why God Won't Go Away*. New York: Ballantine Press, 2001.

Newman, John Henry. *University Sermons*. London: SPCK, 1970.

Noble, William and Iain Davidson. *Human Evolution, Language and Mind: A Psychological and Archaeological Enquiry*. Cambridge, Cambridge University Press, 1996.

Nowak, Martin A. "Five Rules for the Evolution of Cooperation." *Science* 314 (8 December 2006):1560-63.

Oakeshott, Michael. "Education: The Engagement and Its Frustration." In *The Voice of Liberal Learning: Michael Oakeshott on Education*, edited by Timothy Fuller. New Haven and London: Yale University Press, 1989.

Ospovat, Dov. *The Development of Darwin's Theory.* Cambridge: Cambridge University Press, 1979.

—. "God and Natural Selection: the Darwinian Idea of Design." *Journal of the History of Biology* 13 (1980):169-94.

Owen, Richard. *On the Nature of Limbs.* London: Voorst, 1849.

Paley, William. *Evidences of Christianity.* London, 1802.

—. *Natural Theology, or Evidence of the Existence and Attributes of the Deity, collected from the appearances of nature.* Edited by Matthew D. Eddy & David Knight. Oxford: Oxford University Press, 2006.

Paul, Diane B. "Darwinism, Social Darwinism and Eugenics." In *The Cambridge Companion to Darwin,* edited by Jonathan Hodge and Gregory Radick, 214-239. Cambridge: Cambridge University Press, 2003.

Peacocke, Arthur. "Biological Evolution and Christian Theology – Yesterday and Today." In *Darwinism and Divinity*, edited by John R. Durant. Oxford: Blackwell, 1985.

—. *Creation and the World of Science: The Re-Shaping of Belief.* Oxford: Oxford University Press, [1979] 2004.

—. *Evolution, the Disguised Friend of Faith?: Selected Essays*: Philadelphia & London: Templeton Foundation Press, 2004.

Persinger, Michael. *Neuropsychological Bases of God Beliefs.* Westport, Conn.: Praeger, 1987.

Peters, Ted and Martinez J. Hewlett. *Evolution from Creation to New Creation: Conflict, Conversation, and Convergence.* Nashville: Abingdon Press, 2003.

Peters, Ted, and Nathan Hallanger, eds. *God's Action in Nature's World: Essays in Honour of Robert John Russell.* Farnham: Ashgate, 2006.

Phipps, William E. *Darwin's Religious Odyssey.* Harrisburg: Trinity International, 2001.

Polkinghorne, John. "Kenotic Creation and Divine Action." In *The Work of Love,* edited by John Polkinghorne. Grand Rapids: Eerdmans, 2001.

Popper, Karl. "Evolution." *New Scientist*, 87 (21 August 1980): 611.

Rainy, Robert. *Evolution and Theology*. Edinburgh: 1874.

Ramakrishnan, V. "What we have learned from ribosome structures." *Biochem Soc Trans* 36 (Pt 4) (2008): 567-74.

Reidl, Rupert. "A Systems Theory of Evolution." In *Darwinism and Philosophy*, edited by Vittorio Hösle and Christian Illies. Notre Dame, IND: University of Notre Dame Press, 2005.

Relethford, J.H. "Genetic evidence and the modern human origins debate."

Heredity 100 (2008): 555-63.

Rideau, Emile. *Teilhard de Chardin: A Guide to His Thought*. London: Collins, 1967.

Robinson, Andrew, "Continuity, Naturalism and Contingency: A Theology of Evolution Drawing on the Semiotics of C.S. Peirce and Trinitarian Thought," *Zygon* 39 (2004): 111-136.

—. *God and the World of Signs: Trinity, Evolution and the Metaphysical Semiotics of C.S. Peirce* (Leiden, Brill, 2010).

Robinson, Andrew and Christopher Southgate, "A General Definition of Interpretation and Its Application to Origin of Life Research," *Biology and Philosophy* 25 (2010): 163-181.

Robinson, Andrew and Christopher Southgate, "Interpretation and the Origin of Life," *Zygon* 45 (2010): 345-360.

Rodin, A. S., E. Szathmary, and S. N. Rodin. "On origin of genetic code and tRNA before translation." *Biology Direct* 6 (2011): 14.

Romanes, George. *Thoughts on Religion.* London: Longmans, Green and Co., 1896.

Rorem, Paul. *Pseudo-Dionysius: a Commentary on the Texts and an Introduction to Their Influence.* New York & Oxford: Oxford University Press, 1993.

Rosen, Christine. *Preaching Eugenics: Religious Leaders and the American Eugenics Movement.* Oxford: Oxford University Press, 2004.

Rudwick, Martin J.S. "Darwin and Glen Roy: a 'great failure' of scientific method." *Studies in History and Philosophy of Science*, 5 (1974): 97-185.

—. *Bursting the Limits of Time: the Reconstruction of Geohistory in the Age of Revolution*. Chicago: University of Chicago Press, 2005.

Rupke, Nicolaas A. *Richard Owen: Victorian Naturalist.* New Haven: Yale University Press, 1994.

—. *Eminent Lives in Twentieth-Century Science and Religion.* Frankfurt: Peter Lang, 2007.

Ruse, Michael. *Darwin and Design.* Cambridge: Harvard University Press, 2003.

Russell, Colin. "The conflict metaphor and its social origins." *Science & Christian Belief,* 1 (1989): 3-26.

Sandford, E.G. ed. *Memoirs of Archbishop Temple*. London: Macmillan, 1906.

Santmire, H. Paul, and John B. Cobb Jr. "The world of nature according to protestant tradition." In *The Oxford Handbook of Religion and Ecology*, edited by Roger S. Gottlieb. Oxford: Oxford University Press, 2006.

Schloss, Jeffrey P. "Evolutionary theory and religious belief." In *The Oxford Hand book of Religion and Science*, edited by Philip Clayton and Zachary Simpson, 187-206. Oxford: Oxford University Press, 2006.

Schwager, Raymund. *Banished from Eden: Original Sin and Evolutionary Theory in the Drama of Salvation*. Leominster: Gracewing, 2006.

Secord, James A. *Victorian Sensation: The Extraordinary Publication, Reception, and Secret Authorship of Vestiges of the Natural History of Creation.* Chicago: University of Chicago Press, 2000.

Shea, Victor and William Whitla, eds. *Essays and Reviews: the 1860 Text and its Reading*. Charlottesville: University of Virginia Press, 2000.

Sheldon, Joseph K. *Rediscovery of Creation: a Bibliographic Study of the Church's Response to the Environmental Crisis*. Metuchen, NJ: Scarecrow Press, 1992.

Short, T.L., "The Development of Peirce's Theory of Signs," in *The Cambridge Companion to Peirce*, edited by Cheryl Misak, 214-240. Cambridge: Cambridge University Press, 2004.

—. *Peirce's Theory of Signs*. Cambridge: Cambridge University Press, 2007.

Silvertown, Jonathan, ed. *99% Ape*. London: Natural History Museum, 1998.

Souchard, Bertrand. *Aristote de la physique à la métaphysique, réceptivité et causalité*. Dijon: Éditions Universitaires de Dijon, 2003.

Southgate, Christopher. *The Groaning of Creation: God, Evolution and the Problem of Evil*. Louisville: Westminster John Knox Press, 2008.

Spurway, Neil, ed. *Creation and the Abrahamic Faiths*. Newcastle: Cambridge Scholars Publishing, 2008.

—. "What can Evolved Minds Know of God? Reconsidering Theology in the Light of Evolutionary Epistemology." In *Theology, Evolution and the Mind*, edited by Neil Spurway. Newcastle: Cambridge Scholars Publishing, 2009.

—. "Theology's Response to Evolutionary Epistemology. What can Evolved Beings Know of God?" In *Linnaeus and Homo Religiosus. Biological Roots of Religious Awareness and Human Identity*, edited by Carl R. Bråkenheilm. Uppsala: Uppsala Universitet, 2010.

Stamos, David N. "Popper, falsifiability and evolutionary biology." *Biology and Philosophy* 11 (1996): 161-191.

Stearley, Ralph F. "Assessing evidences for the evolution of a human cognitive platform for 'soulish behaviours.'" *Perspectives on Science and Christian Faith*, 61 (2009): 152-74.

Stott, John R.W. *Understanding the Bible.* London: Scripture Union, 1972.

—. *The Message of Romans*. Leicester: IVP, 1994.

Stott, Rebecca. *Darwin and the Barnacle*. London: Faber & Faber, 2003.

Stringer, Chris. "Modern human origins – progress and prospects." *Philosophical Transactions of the Royal Society of London*, B, 357 (2002): 563-79.

Sulloway, Frank J. "Darwin and his finches. The evolution of a legend." *Journal of the History of Biology*, 15: 1-53.

Sumner, J.B. *Evidences of Christianity derived from its Nature and Reception*. J. Hatchard and son: London, 1821.

Symonds, William S. *Old Bones, or Notes for Young Naturalists,* 2nd edition. London: Hardwicke, 1862.

Teilhard de Chardin, Pierre. *Le Milieu Divin: An Essay on the Interior Life*. Translated by Bernard Wall. London: Collins, 1960.

—. *Hymn of the Universe*. Translated by Simon Bartholomew. London: Collins, 1965.

—. *Man's Place in Nature: The Human Zoological Group*. Translated by René Hague. London: Collins, 1966.

—. *Human Energy*. Translated by J. M. Cohen. London: Collins, 1969.

—. *Activation of Energy*. Translated by René Hague. London: Collins, 1970.

—. *Christianity and Evolution*. Translated by René Hague. London: Collins, 1971.

—. *The Phenomenon of Man*. Translated by Bernard Wall. New York: Perennial Library, 1975.

—. *Toward the Future*. Translated by René Hague. London: Collins, 1975.

—. *The Human Phenomenon*. Translated by Sarah Appleton-Weber. Portland: Sussex Academic Press, 1999.

—. *The Future of Man*. Translated by Norman Denny. New York: Image Books/Doubleday, 2004.

Temple, Frederick. *The Present Relations of Science to Religion*. Oxford: Parker, 1860.

—. *Religion and Science*. London: Macmillan, 1884.

Thomson, J. Arthur, and Patrick Geddes. *Evolution*. London: Williams & Norgate, 1912.

Thorpe, W.H. *Purpose in a World of Chance.* Oxford: Oxford University Press, 1978.

Tindal, Matthew. *Christianity as Old as the Creation or the Gospel a Republication of the Religion of Nature*. London, 1731.

Topham, Jonathan. "Science, Religion, and the History of the Book." In *Science and Religion: New Historical Perspectives*, edited by Thomas

Dixon, Geoffrey Cantor and Stephen Pumfrey. Cambridge: Cambridge University Press, 2010.

van der Meer, Jitse. "Theodosius Dobzhansky: Nothing in Evolution Makes Sense Except in the Light of Religion." In *Eminent Lives in Twentieth-Century Science and Religion*, edited by Nicolaas Rupke. Frankfurt: Peter Lang, 2007.

van Huyssteen, J. Wenztel. *Alone in the World? Human Uniqueness in Science and Theology*. Grand Rapids: Eerdmans, 2006.

van Till, Howard. "The Creation: Intelligently Designed or Optimally Equipped?" *Theology Today* 55 (1998/9): 344–364.

von Balthasar, Hans Urs. *The Glory of the Lord: A Theological Aesthetics* Vol VII: Theology – The New Covenant. San Francisco: Ignatius Press, 1989; German original 1969.

Wallace, Alfred Russel, "The origin of human races and the antiquity of man deduced from the theory of natural selection," *Journal of the Anthropological Society of Lond on* 2 (1864): 157-187.

—. *My Life,* new edition. London: Chapman & Hall, 1908.

Wallace, Ian, ed. *Leonard Jenyns: Darwin's Lifelong Friend*. Bath: Royal Literary and Scientific Institution, 2005.

Warfield, B.B. "On the antiquity and unity of the human race," *Princeton Theological Review* 9 (1911): 1-25.

Westermann, Claus. *The Genesis Accounts of Creation*. Philadelphia: Fortress, 1964.

—. *Genesis 1-11*. Translated by J.J. Scullion. London: SPCK, 1984.

White, Lynn. "The Historical Roots of Our Ecologic Crisis." *Science* 155 (1967): 1203-07.

Whitehead, Alfred N. *Process and Reality, an Essay in Cosmology*. Corrected Edition, edited by David Ray Griffin and Donald W. Sherburne. New York, London: The Free Press, [1929] 1978.

Wilberforce, Samuel. "Darwin's Origin of Species." In *Essays Contributed to the Quarterly Review*, 2 vols. London: Murray, 1874.

Williams, Patricia A. *Doing without Adam and Eve*. Minneapolis: Fortress Press, 2001.

Wilson, David Sloan. *Darwin's Cathedral: Evolution, Religion and the Nature of Society*. Chicago: Chicago University Press, 2002.

Wilson, Edward O. *Sociobiology*. Cambridge: Harvard University Press, 1975.

—. *Consilience.* New York: Knopf, 1998.

Wochner, A., J. Attwater, A. Coulson, and P. Holliger. "Ribozyme-catalyzed transcription of an active ribozyme." *Science* 332 (6026) (2011): 209-12.

Woit, Peter. *Not Even Wrong: the Failure of String Theory and the Search for Unity in Physical Law*. New York: Basic Books, 2006.

Wolterstorff, Nicholas . "The Migration of the Theistic Arguments: From Natural Theology to Evidentialist Apologetics." In *Rationality, Religious Belief and Moral Commitment*, edited by Robert Audi and William J. Wainwright. Ithaca: Cornell University Press, 1986.

Wright, Sewall. "Evolution in Mendelian populations." *Genetics*, 16 (1931): 97-159.

Wynne-Edwards, V.C. *Animal Dispersion in Relation to Social Behaviour*. Edinburgh: Oliver & Boyd, 1962.

Yarus, M., J. G. Caporaso, and R. Knight. "Origins of the genetic code: the escaped triplet theory." *Annual Review of Biochemistry* 74 (2005):179-98.

Zeldovich, K.B., P. Chen, B.E. Shakhnovich, and E.I. Shakhnovich. "A first-principles model of early evolution: emergence of gene families, species, and preferred protein folds." *PLoS Comput Biol* 3 (7) (2007): e139.

Zizioulas, John. "Creation theology: an Orthodox perspective." In *Listening to Creation Groaning,* edited by L. Vischer, 90-104. Geneva: John Knox International Reformed Center, 2005.

—. "Priest of creation." In *Environmental Stewardship*, edited by R.J. Berry, 253-60. London: T&T Clark, 2006.

INDEX